THE INSHORE FISHERMEN
OF WALES

THE INSHORE FISHERMEN OF WALES

J. GERAINT JENKINS

AMBERLEY

First published 1991
This edition published 2009

Amberley Publishing
Cirencester Road, Chalford,
Stroud, Gloucestershire, GL6 8PE

www.amberley-books.com

British Library Cataloguing in Publication Data.
A catalogue record for this book is available from the British Library.

ISBN 978 1 84868 158 3

Typesetting and Origination by Diagraf (www.diagraf.net)
Printed in Great Britain

CONTENTS

PUBLISHER'S NOTE

This book was first published in 1991. The book has been re-typeset with minor corrections by Dr J. Geraint Jenkins, but essentially the text remains the same as the 1991 edition. The reader should be aware of this in relation to chronological references.

PREFACE

Over the centuries the inhabitants of the coastal areas of Wales have exploited the harvest of the seas and in many districts fish formed the only item of export. The survival of many a coastal community depended on a flourishing fishing industry.

In Wales, as elsewhere, the fishing industry is in a state of flux and the picture that emerges is an ever-changing one. New, international restrictions on fishing grounds and quantities of fish caught, ever-increasing costs, constantly changing regulations as well as the introduction of new techniques and modern equipment have all had their effect on the industry. In recent years the unchanging ways of fishermen and fisherwomen that could be traced back for a thousand years or more have been revolutionized and this book sets out to record and present the wisdom of centuries in the coastal communities of Wales.

In preparing this volume I would like to thank my colleagues Lynn Davies, Arwyn Lloyd Hughes, Mrs Minwel Tibbott, Dr David Jenkins, Mrs S.M. Charles and Mrs Anne Bunford for their help and support. R.J.H. Lloyd of Pendoylan who has tremendous knowledge of traditional coastal craft supplied me with a great deal of information and inspiration. I owe a great debt of gratitude to the late Colin Matheson, one-time keeper of Zoology at the National Museum of Wales who carried out a pioneering study of Welsh coastal fishing in the 1920s. Above all, I am grateful to the fishermen and fisherwomen of Wales, the trawler owners and the staff of the various fishing authorities for their time and forbearance over the course of many years of sporadic field-work.

A NOTE ON ILLUSTRATIONS

Unless otherwise acknowledged, all photographs in this book are from the archives of the National Museum of Wales.

INTRODUCTION

Wales has a long, indented coastline of almost a thousand miles. It has broad river estuaries, numerous bays and peninsulas and many off-shore islands. Due to its geography it was almost inevitable that Wales should have developed a maritime tradition that may be traced back to prehistoric times. An important element of this heritage was the fishing industry, for inshore and estuarine fishing as well as the gathering of shellfish have always been of considerable importance in the economic life of the Principality. Fishing provided a livelihood for many generations of riverside and coastal dwellers and it is only natural that those village communities where fishing was of importance should have developed their own character and personality. Many of the coastal settlements in particular were, in effect, self-contained communities that looked outwards towards the open sea. The oceans dominated the lives and thinking of the inhabitants of those settlements. By tradition those coastal villages had little contact with the surrounding countryside and there was a sharp dichotomy between the inhabitants of the outward-looking maritime communities and the inward-looking agricultural communities of the adjacent countryside. For example the coastal villages of Ceredigion; villages such as Aber-porth, Llangrannog and Llanddewi Aber-arth were self-contained communities and their inhabitants had no outlook but the Irish Sea. In those villages, the capture of the herring was all-important and the capture of that fish supplied a vital element in the diet of the people, providing a livelihood for many generations of coastal dwellers. By the late seventeenth century, the herring of Cardigan Bay had become much more than 'a survival food', for it became almost the sole item of export from an area that had few goods to trade. The basis of the intense commercial activity that the coastal villages were to witness in the eighteenth and nineteenth centuries was the capture and export of the herring found in such great quantities in Cardigan Bay.[1] It was the presence of the herring shoals that attracted both merchants and mariners to the coastal villages of Ceredigion and gave rise to the rapid growth of the villages themselves.

By the outbreak of the Great War in 1914, herring fishing in Wales had declined together with the wider commercial maritime activity it brought into being, but in more recent times a few coastal villages whose whole economy has been tied up with the fishing industry have survived. One such village is that of Pen-clawdd on the north Gower coast together with the adjacent hamlets of Croffty and Llanmorlais. Here cockle gathering remained the main occupation of the female inhabitants although since 1978 there has been a sharp decline in the number of gatherers. In 1970 there were as many as ninety-two licensed cockle gatherers in the three villages, who day in, day out travelled the three miles of treacherous shifting sands to gather cockles on Llanrhidian beach. North Gower villages with their dependence on shellfish-gathering from the shores of the Burry Inlet were unique communities where the female population dominated both social life and economic welfare.

On a much larger scale the fishing industry was all-important in such places as Milford Haven and in Conwy, Moelfre, Aberdaron and Barmouth to a lesser degree. Although Milford Haven, for example, had ambitions to be a large trans-Atlantic port the Dock Company did not build 'their great docks to attract fishing smacks' they nevertheless 'did not look askance at the great possible development of the fish trade'[2] By 1912 when the Atlantic dream had ended, Milford was a port where the fish trade was the only industry. 'Everything and everybody depend on it. Directly or indirectly between 1500 and 2000 people are engaged in it. The population of the town has been doubled by means of it and thousands of pounds worth of house property has been erected as an outcome of its prosperity.'[3]

The strong sense of community so characteristic of fishing groups was by no means limited to coastal villages and towns, for it was also found amongst riverside dwellers as well. The coracle fishermen of west Wales, for example, were described by the Commissioners on Salmon Fishing in 1861 as 'a numerous class, bound together by a strong *esprit de corps* and from long and undisturbed enjoyment of their peculiar mode of fishing, have come to look on their river almost as their own and to regard with extreme jealousy any interference with what they consider their rights'.[4] The salmon fishermen of the four riverside villages of Llechryd, Cenarth, Abercuch and Cilgerran on the river Teifi formed almost closed communities and custom had dictated that only fishermen from these four villages had the right to fish for salmon in their particular reaches of the Teifi.[5] There were rigid rules of privilege and procedure that were passed down from father to son in the closely knit communities that depended very largely on the harvest of the river. In that section of the Teifi in the early nineteenth century 'there (was) scarcely a cottage in the neighbourhood without its coracle hanging by the door'.[6]

The drift-netsmen of the Dee estuary (the majority being members of the same family) and the seine-netsmen concerned with salmon capture in many of the river estuaries of Wales, were distinct and close-knit groups that always displayed a sense of community. In many groups it was almost impossible for strangers to enter the fraternity of fishermen who administered a river through set rules of privilege and rights that had been passed down from father to son from time immemorial. Most were oral laws that have persisted almost to our own day despite the many economic, social and cultural changes in riverside communities.

1 THE FISH

The rivers and the seas around Wales are particularly rich in fish and the fishing industry has provided a livelihood for many generations of Welsh people. George Owen of Henllys, for example, writing in 1603, speaks of the great variety of fish that occurred in Pembrokeshire providing 'one of the cheefest worldlie commodities, wherewithal god hath blessed this countrye'.[1] Salmon and turbot, oysters and lobsters, herring and eels were taken in considerable quantities and fishing was undoubtedly one of the main pursuits of south-west Wales in George Owen's day. The economy of Bardsey Island, to quote another example, was based very firmly on the capture of the variety of fish that occurred around the island while villages and creeks on the adjacent Llŷn mainland supported numerous inshore fishermen. The economy of such coastal towns as Aberystwyth and Pwllheli, Tenby and Barmouth was largely dependent on fishing fleets and in recent times, ports such as Milford Haven and Conwy saw periods of considerable prosperity as major fishing ports.

This book is primarily concerned with commercial fishing in the river estuaries of Wales, the gathering of shellfish along the coasts and inshore fishing in the seas around the Principality. In the past a very wide variety of fish was caught, but the following were the principal species.

River fish

SALMON (*Salmo salar*); Welsh Eog[2]

Almost all the commercial fishing in the estuaries of Wales is traditionally associated with the capture of salmon and its close relative the migratory trout or sewin. The fish, still caught in considerable quantities in the river estuaries of Wales, has been caught for many centuries with a wide range of fixed or movable equipment.

The salmon is a large delicately flavoured fish found in seas and rivers both sides of the Atlantic. Adult salmon ascend the rivers, mainly in the summer months, to the gravelly shallows or *redds* (Welsh *maran*, *maranedd, clâdd* or *gwely*) where breeding takes place between September and February. The salmon does not feed at all in fresh water and, as a result, by the time they reach the spawning grounds, they are in poor condition.

A hen salmon (Welsh *hwyfen, hwyddell* or *chwiwell*) lays between eight and nine hundred eggs for every pound of her weight and the eggs are fertilized by the cock salmon (Welsh *cenyw*) before being covered by gravel. In the spring the eggs are hatched and the tiny salmon or *alevins* (Welsh *silod*) are provided with a yolk sac which is gradually absorbed by the growing fish. The adult salmon after spawning return to the sea as *kelts*; many will die before they reach the sea; a few will return to the same spawning grounds in later years.

Meanwhile, the alevins grow and begin to feed in the river. The alevin becomes a *parr* (Welsh *eginyn*), a trout-like fish with a series of dusky mauve marks along its sides. As a parr it remains in the river for a year or more and gradually changes its colour to silver. The parrs then congregate in considerable numbers, when they are known as *smolts* and they move towards the sea, where the major part of the growing and feeding is done. They sometimes return to the same river as they were born after a year at sea and the *grilse*, as the year-old salmon is termed, weighs anything from four to nine pounds. Many will not return for two or more years and the fish are consequently much heavier and larger than the grilse. The Wye is particularly famous for its large-sized salmon, many of them not returning to the river until they have spent four years away at sea. The surviving salmon always return to their river of birth.

'Salmon fishing' said Abraham Rees in 1819 was 'of great national importance furnishing a constant and copious source and supply of human food. The fisheries may be said with propriety to rank next to that of the cultivation of the land in utilising this intention'.[3] For many centuries, Welsh rivers have been renowned for the quality and quantity of their salmon. For example, Giraldus Cambrensis in the twelfth century described that notable salmon river, the Teifi, as 'the noble river Teivi that abounds more than any river of Wales with the finest salmon'.[4] George Owen spoke of 'Ryver fishe, whereof the saman shall have the first place, partely for the plentie and store thereof.'[5] Travellers to Wales in later centuries were all impressed with the quality and abundance of salmon in Welsh rivers and today the fish is by far the most important in value taken from the rivers of the Principality.

SEA TROUT (Sewin: Migratory Trout), (*Salmo trutta*), Welsh *Gwyniedyn*, pl. *Gwyniaid*

The sewin or sea trout has a similar life cycle to the salmon and its flesh is similar although usually lighter in colour. In some rivers, notably the Tywi, Dyfi and Conwy, the netting of sewin is as important, if not more important than fishing for salmon. The Tywi is noted particularly as a sewin river; so much so that specific names are used to indicate the weight of fish caught. A sewin weighing 3 to 20 pounds is called a *gwencyn*; that between 2 and 3 pounds is called a *twlpyn*, while those less that 2 pounds are called *shinglin*. Oddly enough, there are many rivers in Wales where negligible quantities of sewin are caught. Although they occur on the Usk, for example, sewin are almost unknown on the adjacent Wye. None is caught on the Dee, Clwyd or Severn.

EELS (*Anguilla anguilla*), Welsh *Llysywen*, pl. *Llyswennod*

Eels occur widely throughout western Europe and they are found in rivers and inland waterways throughout Britain. In whatever stretch of water eels are found it is certain that they have travelled from their breeding ground

Elver fishing in the Severn estuary (Photo – City of Gloucester Museum)

in the Sargasso Sea, not far from Bermuda and the Leeward Islands, where they breed at a depth of between 500 and 600 fathoms. It is in the Sargasso Sea alone that young eel larvae have been found and it is to this area too that adult eels will return to breed and die.

Immediate after hatching, the larvae in their millions begin their long journey eastwards to western Europe, a journey that takes up to three years. The larvae, know as *leptocephalids*, are tiny transparent bodies, but as they near the European coast in the autumn of their third year of life, they change into young eels or *elvers*, each about 2 or 3 inches long. In the lower Severn Valley, below Gloucester, the elvers are caught in great quantities and fried elver is regarded as a great delicacy on the banks of the lower Severn.

Most of the elvers that enter British waters, however, are not caught but spread upstream to canals, brooks, ditches and they even reach isolated ponds. In these places they feed and grow until they reach maturity, between eight and twelve years of age.

Eels that are feeding and growing are known as *yellow eels* or *gelps* and are green or yellow on their undersides. Although yellow eels were often caught in the past, most of those caught today are silver eels, eels that change their colour to silver when they are about to migrate to the sea. Migration takes place in autumn, especially on dark nights when the river is in flood.

Not many eels are caught in Wales nowadays, but a certain amount of netting, spearing and trapping is carried out in the Severn estuary area. The only fixed eel trap in use today is at Llangorse Lake (Llyn Safaddan) in south Powys, where a trap is in operation during the autumn months.

ALLIS SHAD (*Alosa alosa*) and TWAITE SHAD (*Alosa fallax*)

The shad, a migratory fish, ascends the larger rivers in shoals in late spring and early summer to breed in fresh water. A close relative of the herring, it is the same colour – silver with a blue-green back and is up to two feet in length. The species live in British waters; the twaite shad which enters the rivers in late May and June and the larger allis shad. which arrives about a month earlier. The shad is a very bony fish and it was a common practice to net them in the Severn estuary although few have been caught within the last hundred years. A few are still netted on the Wye at Symonds Yat, but the fish that needs considerable preparation is no longer popular amongst riverside dwellers. 'In former days shad, probably the larger allis shad, were quite highly esteemed and Henry III frequently received supplies from the Severn, particularly during Lent.'[6]

LAMPREY (*Petromyzon marimus*) and LAMPERN (*Lampetra fluviatilis*)

The large sea lamprey, a stone-sucker, enters the rivers of Wales, especially the Severn, early in the year to breed in fresh water. Spawning takes place in shallow water in April and May. The young fish remain in fresh water for from three to five years, then they undergo metamorphosis and develop sharp horny teeth. At this stage they migrate to the sea and become semi-parasitic, preying on other fish. Two years later they return to fresh water again to spawn. Although lamprey pie was said to be a delicacy in the Middle Ages, the revolting-looking fish is now despised as a parasite. On the Tywi, for example, the coracle men of Carmarthen are unwilling to use fine-meshed nets for the capture of salmon before the main run of lampreys is over about mid-June.

The lampern or river lamprey is smaller than the sea lamprey and has a similar life history but they may run up river at any time between September and April, spawning in May and June.

Despite their popularity with medieval kings, very few lampreys are caught today – the few that are taken are used as eel bait. In the nineteenth century considerable quantities of Severn lampreys were sent to east-coast ports for use as cod bait.

SPARLINGS (*Osmerus eperlanus*), Welsh *Brwyniad*, pl. *Brwyniaid*

The sparling occurs in the rivers of north-east Wales but is today caught in the Conwy only. Considerable folklore surrounds this cucumber-smelling fish, for local tradition states that sparlings occur on no river but the Conwy and they do not appear until all the snow on the mountain peaks has disappeared. The fish is connected by legend with Saint Brigid or Sant Ffraid, who sailed from Ireland on a piece of detached ground and landed on the foreshore at Glan Conwy, where the ground became part of the coast on which the saint built a church. Legend says that when there was great famine in north Wales, St Brigid threw a handful of rushes (*brwyn*) into the river and they turned into sparlings (*brwyniaid*), which provided food for the starving population.

Among the other fish that enter estuaries are bass (*Dicentrarchus labrax*), and the thick-lipped mullet (*Chelan labrosus*) but these are not of great importance. There are other marine and estuarine species that frequently enter fresh water for short periods. Some flounders are caught in Burry Inlet in south Wales and in the Dee estuary.

Shellfish

LOBSTER (*Homarus gammarus*), Welsh *Cimwch*, pl. *Cimychiaid*

The lobster is Britain's largest crustacean and specimens as heavy as 14½ pounds have been caught off the south-west Dyfed coastline. The average weighs 2 to 3 pounds and they normally measure from 8 to 20 inches in length. The fish, which lives along the rocky sections of the Welsh coast, is blue with brown markings when alive but turns red when cooked. Today, lobsters are by far the most important of the shellfish exported from Wales and appreciable quantities are caught off the rocky coasts of Dyfed, Llŷn and north and west Anglesey. Substantial quantities of live lobster are exported to the European continent, particularly to France.

CRAB (*Cancer pagurus*), Welsh *Cranc*, pl. *Crancod*

A crab which measures up to 10 inches across the shell and can weigh as much as 12 pounds lives some distance off-shore and like the lobster is caught in wickerwork, wire or plastic pots. The Cardigan Bay ports of New Quay, Barmouth and Aberdaron are important for crab fishing as is the north coast of Anglesey and St David's Head in Dyfed.

CRAWFISH (*Palinurus vulgaris*), Welsh *Cimwch coch*

The crawfish, also known as the spiny lobster or rock lobster is slightly larger than the true lobster, but has small pincers. Like lobster they are found on rocky grounds off the west coast. Most of those caught by fishermen from such ports as New Quay and Barmouth are exported to France where the crawfish is know as langouste. Until recently few crawfish were caught by Welsh fishermen, but Breton fishermen are now a common sight along the coasts of Pembrokeshire, in particular searching for the catch that is regarded as a great delicacy in France.

MUSSELS (*Mytilus edulis*), Welsh *Cregyn Gleision*

Mussels are found all round the coast of Wales wherever there are suitable rocks to which they can attach themselves. Those which live high up the shore remain small, since they feed only when covered by water. Commercial

Mussel fishermen unloading the catch at Conwy in the 1930s

exploitation of the common mussel is concentrated in river estuaries. Up until the late 1970s the most important fishery of all was in the Conwy estuary where mussels were gathered either on the shore (*o'r lan*) or from the deep water (*o'r dwfn*) by licensed fishermen in the winter months. Most of the mussels are sold in the markets of the north of England. Until about 1980 mussels were also gathered at the mouth of the River Glaslyn and Porthmadog harbour, while the farming of mussels in the Menai Straits is of increasing importance, and a certain amount are also gathered in the Burry Inlet of Carmarthen Bay. The best mussel beds occur below the low-water mark where the mussels are continuously covered by water and can feed constantly in water that is full of organic material carried down by the rivers.

COCKLES (*Cerastoderme [cardium] edule*), Welsh *Cocos, Cocs, Rhython*

Edible cockles live close to the surface of the sand on sheltered tidal shores, the most extensive beds occurring in Carmarthen Bay, especially the Burry Inlet. As cockles draw in sand with the plant plankton they feed on, they must be kept for some hours in the sea water before being cooked, so that they have time to expel the sand. In the Burry Inlet most of the cockle gathering is carried out on Llanrhidian Sands by fishermen and women from three north Gower villages – Pen-clawdd, Croffty and Llanmorlais.

SHRIMPS (*Crangon crangon*), Welsh *Perdys* and PRAWNS (*Palaemon serratus*), Welsh *Corgimwch*, pl. *Corgimychiaid*

The shrimp is up to 3 inches in length and has a flat body which can change colour from yellow to almost black to match the colour of the sand over which it lives. It occurs widely on many parts of the Welsh coast. Some are caught with trawl nets and traps in the Severn, Wye and Dee estuaries. The prawn is abundant, particularly in autumn all around Wales. Its body increasing from 1 to 4 inches in length is transparent and is dotted with purple pigment cells. The much larger Dublin Bay prawn or scampi (*Nephrops norvegicus*) is not a true prawn but a close relative of the common lobster. Only a few are caught off the Welsh coast.

OYSTER (*Ostrea edulis*), Welsh *Wystrys*

The edible oyster, found off-shore in shallow water, was until the beginning of the twentieth century of considerable economic importance in Swansea Bay and to a lesser extent off the south Pembrokeshire coast. Despite the fact that oyster beds have been artificially planted in parts of Wales, oyster dredging has virtually disappeared from the Principality. Nevertheless a flourishing oyster fishery was established in the calm waters of Milford Haven in the 1980s and has proved very promising.

SCALLOPS (*Pecten maximus*) and QUEEN SCALLOPS (*Chlamys opercularis*)

Scallops live in deep water off-shore on sands or gravelly sea beds. The fish is dredged commercially in Cardigan Bay and considerable quantities were, until 1985, treated in a processing plant in the Ceredigion village of New Quay.

LIMPETS (*Patella vulgata*), Welsh *Llygaid Meheryn, Llygaid y Craig*

Limpets, which are rock-clinging snails with conical shells, occur widely on the sea-shore. They were never gathered commercially but were a survival food amongst the coastal communities of Ceredigion and Llŷn in particular.

Usually they were boiled and the flesh then fried in bacon fat. With the addition of bacon and a leek, limpet pie (*tarten lygaid*) was a favourite food in the Aber-porth district of Ceredigion.

PERIWINKLES (*Littorina littorea*), Welsh *Gwichiaid*

A brown-black snail common on rocky shores was also a survival food gathered from September to April especially along the rocky shore of the Llŷn peninsula. They were covered with salty water and brought to boiling point, then allowed to cool in the water before being removed from the shells and eaten. On Bardsey, winkles were fried in bacon fat and mixed with an egg to make an omelette.

Among other fish gathered from the sea-shore were sand eels (*Ammodytes tobianus*), Welsh *Llymriaid*, thin fishes up to 8 inches in length that were found on beaches in west Wales in the summer months. On Penbryn Beach in Ceredigion, *llymreita* or the gathering of sand eels was carried out in moonlight after the tide had receded, especially in the month of August. A shovel or sickle was needed to dig in the sand and the tiny eels were gathered in buckets. With the heads removed and the insides squeezed out the eels were fried in bacon fat.

Sea fish

The seas around Wales have an abundance of fish, although the sea fishing industry today is far less important than it was fifty years ago. Some species of fish, such as the hake, on which the fortunes of the Milford Haven fishing industry were based, have become much rarer in Welsh waters, while herring fishing, that brought prosperity to many a coastal village over the course of several centuries, has become almost insignificant.

There are two major classes of fish that occur in Welsh waters – pelagic fish and demersal fish. Pelagic fish include such species as herring, mackerel and pollack which live and feed near the surface of the water. Their capture with drift nets in particular, has always been the main preoccupation of inshore fishermen who use small boats from the coastal settlements of west and north Wales. Demersal fish on the other hand, live for the most part on or near the bottom and include such species as hake, cod, plaice and haddock. Most of the demersal fish were captured by trawl nets operated by fishing vessels from the more important Welsh ports such as Milford Haven, Swansea and Cardiff.

Along the north coast of Wales there is fish in plenty although commercial

The Milford Haven fleet of steam trawlers alongside the fish market in the 1920s

fishing never developed to the extent that it should for much of the industry was in the hands of fishermen from Fleetwood and the Isle of Man. Flat-fish such as plaice, turbot, flounder, brill, sole and dab; roundfish such as cod, whiting, mackerel, herring and mullet occur profusely – the sands and rocky inlets round the coast of Anglesey in particular were traditionally areas of considerable richness.

In Cardigan Bay, which is shallow and fed by a large number of rivers, a variety of fish occur. Cardigan Bay is also liable to be affected by tidal currents and by disturbance from the Atlantic. Whiting, bass, plaice, turbot, brill, sole and hake are all found in the Irish Sea while from time to time great shoals of herring and mackerel enter the Bay and in the past, at least, their appearance near the shore ushered in a period of unprecedented activity. The deeper waters of the Irish Sea were worked by trawlers from Milford, Fleetwood and Irish ports.

The Bristol Channel lying open to the Atlantic is well supplied with a great variety of fish, although the commercial exploitation of the fish has substantially declined in recent years. The extensive Cardiff fishing fleet has virtually disappeared and Swansea is no longer the major fishing port that it was in the 1920s and '30s. Before the First World War the Bristol Channel attracted about a hundred Brixham smacks every summer which worked from Milford. Herring, mackerel, hake, cod, ling, whiting, pollack, bream, conger eels, skate, ray, sole and many other varieties occurred profusely. 'The Bristol Channel', wrote Matheson in 1929,[7] 'has been described as "the home of the sole" and its eggs occur there in great quantities ... "Barry soles" have long been famed for their delicacy and flavour.' Oddly enough two important food fishes – the haddock and halibut – have never been common off the Welsh coast and those landed at such places as Milford Haven in the inter-war period were caught by deep-sea trawlers operating in the Atlantic.

HERRING (*Clupea harengus*), Welsh *Penwaig, Ysgadan*

In the past herring was extensively fished round the coast of Wales and inshore herring fishing was the main occupation of villages along the shores of Cardigan Bay. Within the last eighty years there has been a sharp decline in herring fishing in Wales and it is said that the vast stocks of fish no longer come close to the shore in Cardigan Bay. In the heyday of herring fishing, the capture of the fish usually took place between late August and early February. During those months the herring swam close to the shore and near the surface of the water and could be caught with drift nets. Before 1939, Milford was the principal station in Britain for landing herring. The Irish Sea herring is an autumn-spawning fish in contrast to the North Sea, spring-spawning variety. Until the end of the nineteenth century, Nefyn in Gwynedd and Aber-porth in Dyfed were important herring fishing ports.

MACKEREL (*Scomber scomber*), Welsh *Mecryll*

Shoals of mackerel appear along the Welsh coast in the summer months and although they may be netted as they often are today, the traditional method of capturing them was with long lines of baited hooks. The mackerel is a migratory fish that comes close inshore in vast shoals for spawning and then returns in smaller shoals to the deep water of the North Atlantic. Today, vast quantities of mackerel are landed by freezer trawlers at Milford Haven throughout the year and the fish is exported principally to West Africa. The modern trawlers employed in this profitable trade are not Welsh vessels and most of them operate in deep water.

HAKE (*Merluccius vulgaris*), Welsh *Cegddu*

Much of the prosperity of Milford Haven as a fishing port was based on the capture of hake off the south-eastern coast of Ireland. The over-fishing of those once profitable grounds by Spanish and French fishermen led to the rapid destruction of the hake grounds, and the consequent decline of Milford Haven as one of the most important fishing ports of Europe. In the heyday of hake fishing, Welsh trawlers not only fished the Irish Sea but wandered further afield to the west of Scotland and Ireland and even as far as Portugal, while appreciable quantities were caught by Cardiff and Swansea trawlers in the Bristol Channel.

A great variety of other sea fish was caught by fishermen in the waters around Wales. Some were strictly for local consumption, others were sold outside the area. Thus in the Llŷn peninsula, the fishermen of Aberdaron always regarded the lobster as a saleable fish and very few were consumed within the community itself. A few crabs were eaten as was herring, either fresh or salted or smoked. In the 1930s off the Llŷn coast many different species of fish were caught.[8] Amongst them was the occasional tope (*ci glas*), skate (*cath fôr*), bass (*draenogiad*) cod (*penfras*), pollack (*gwrachen y môr*), sole (*lleden chwithig* or *tafod yr hydd*) and ling (*honos* or *breninbysg*). Some of these were cut up as bait for lobster pots, others were salted and dried for consumption in the winter.

In Milford Haven, after the decline of the home fisheries in the 1950s, the few remaining trawlers that operated from the port in the 1960s and 70s fished the Irish Sea and Bristol Channel grounds and the most important fish landed during that period of decline was the roker or thornback, a member of the skate family. In 1963, 41,465 cwt. of roker was landed at Milford. This was followed by cod (24,769 cwt.), whiting (19,846 cwt.) and haddock (15,390 cwt.). Minor quantities of conger eel, dog fish, ling, plaice, mullet and sole were also landed and a few hake from the Scottish coast were also being brought in although that fishery was to decline soon after. In recent years the sea bass in particular has become an important and sought-after fish in all Welsh waters for it commands a high price in the market.

2 SALMON FISHING

The salmon is by far the most sought-after fish in the river estuaries of Wales and most of the commercial fishing and certainly all the illegal poaching carried out in the Principality is concerned with the capture of 'the King of fishes'. Some of the methods of fishing may be traced back for centuries and many of the instruments used have persisted in Wales and its eastern borders whereas they have completely disappeared from other regions. Thus in the tidal reaches of three west Wales rivers – the Tywi, Taf and Teifi – the coracle, which may be traced back to prehistoric times, is still in regular use for salmon capture.[1] Old established fish traps of willow still persist on the shores of the Bristol Channel with its stupendous rise and fall of tides. The complex trammel-net is still used by members of one fishing family in the Dee estuary.

Nevertheless all is not well in the commercial salmon-fishing industry in Wales and since 1980 a number of traditional methods have ceased to exist. Stop-net fishing, where large V-shaped nets were suspended from moored boats on the Wye at Chepstow and at Wellhouse Bay in the Severn estuary, ceased in 1985, after hundreds of years of operation. Licences on putcher ranks in Gwent were not taken up and netting seasons on such rivers as the Cleddau were severely reduced.

The following were licensed to fish for salmon on a commercial basis in the 1980s:

River/Area	Instrument	1982	1985	1989
Wye	Drift-nets	2	-	-
	Stop-nets	6	-	-
	Lave-nets	10	9	8
Usk	Drift-nets	8	8	8
	Putcher ranks	3	3	-
			(not operated)	
Tywi	Coracle-nets	12	11	12
	Seine-nets	9	9	9
Taf	Coracle-nets	2	1	1

	Wade-nets	1	1	1
Cleddau	Compass (Stop) Nets	8	8	8
Nevern	Seine-net	2	2	1
Teifi	Coracle-nets	15	12	12
	Seine-nets	6	6	6
Dyfi	Seine-nets	6	6	6
Dysynni	" "	2	1	1
Mawddach	" "	3	3	3
Dwyryd & Glaslyn	" "	3	2	2
Dwyfor	" "	2	2	2
Daron	" "	2	2	2
N. Caernarfon	" "	2	2	2
Seiont & Menai	" "	5	4	4
N. Anglesey	" "	2	2	2
Ogwen	" "	2	2	2
Conwy	Fishing weir	1	1	1
	Seine-nets	6	6	6
	Basket trap	1	1	1
Clwyd	Drift-nets (sling-nets)	8	8	8
Dee	Seine-nets	29	30	30
	Trammel-nets	4	4	4

Hand-netting

In the past small, hand-operated nets for use by a single fisherman were widely employed in many river estuaries and along the coast. Shallow water in which a man could wade and move was essential, for in hand-netting rapid movement to catch a moving fish as it made its way upstream in the shallows was vital and the use of such nets demanded a high degree of skill and agility. On the banks of the Severn very fine meshed nets attached to a handstaff are still used to catch the tiny elvers that make their way up river in the spring months[2] but as far as the salmon is concerned, the best-known of hand-nets is the so-called lave-net. This is still used in the shallows of the Severn estuary and consists of a Y-shaped frame with a net attached, used at low water. It is an ancient instrument described in some detail as a 'becknet' or 'ladenet' in the early seventeenth century.[3]

To use the lave-net the fisherman stands in shallow waters, with the net at the ready. A fishing expedition may involve a fruitless wait of many hours. As soon as the wake of a salmon rushing upstream through the shallows is seen, the netsman runs to intercept its path. He lowers the net into the water just as the salmon approaches and once it is over the headline and in the net, the fisherman has to step back smartly to counteract the force of the rapidly moving fish. The hand staff is raised, flipped over and the salmon killed with a wooden knocker or 'priest'. This tool, an essential for all salmon fishermen, is of apple or holly wood about 12 inches long, usually attached to a lanyard round the fisherman's waist. Lave-netting is a most dangerous method of fishing and lave-netsmen have to keep an eye on the tide as well as the fish. Flooding in the Severn estuary is rapid and sudden and often fishermen work in pairs, so that while one man fishes, the other keeps a look out for the incoming tide.

Stop-net fishing

Stop-net fishing involves the use of a large bag-like net suspended from two heavy poles that form a V-shaped frame. Fishing is done against the tide from boats that take up station against poles driven into the bed of the river, or moored to a steel cable fixed to both shores of a river. Stop-nets were used in three places: 1) on the River Severn at Wellhouse Bay in Gloucestershire, a few miles from the Welsh border; 2) on the River Wye at Chepstow; 3) on the eastern and western Cleddau in Milford Haven in Dyfed. Here the nets are called 'compass nets' due to their resemblance to a pair of compasses.

In 1984, stop-net fishing at Wellhouse Bay ceased and the broad beamed 35-foot vessels traditionally used for fishing were idle for the first time in centuries. When in use, the boats were moored to a steel hawser called 'the chain' that extended for about 200 yards from the river bank. In the heyday of stop-net fishing, there were seven of these chains and the six boats that were utilized in the 1960s could be tied anywhere along the length of these seven chains.

In 1985 stop-netting on the Wye at Chepstow also ceased and with the closure of the fishery and its associated buildings a tradition that may be traced back for hundreds of years disappeared. The heavy, Chepstow-built stopping boats, each one manned by a single fisherman, were anchored

Stop-net fishing on the river Wye above Chepstow: the waiting game with the stop-net in fishing position against the flow of the tide

across the flow of the river on the ebbing tide. In the early 1970s as many as a dozen of these grey and black, broad-beamed 25-foot long boats were in constant use below Chepstow Bridge and in the 1940s a further seven boats fished near Beachley. In the nineteenth century there were as many as 37 stopping boats in constant use and these went up river as far as Brockweir bridge and some even at Tintern. Unlike the Wellhouse Bay boats that were attached to the shore by chains, the Wye boats were anchored in mid-stream and a boat was kept in position by three 32-foot wooden stakes driven into the bed of the river. Since stakes were used for mooring, the fishing session was limited to no more that three hours when the depth of water was not too much for the efficient moving of the boats.

On the eastern and western Cleddau in Dyfed, stop-nets were termed 'compass nets' and the method of fishing is said to have been introduced into Dyfed by the Gloucestershire men, Osmond and Edwards, who came to work in the Landshipping coal mine in the late eighteenth century. Two fishing stations on the eastern Cleddau still bear the names of Osmond and Edwards. The nets are operated by eight licensees from the villages of Hook and Llangwm, the Hook fishermen being limited to the western Cleddau only

Stop-net fishing on the river Wye: 'knocking out'.

but the Llangwm fishermen being allowed to fish in the two rivers. Ordinary 14-foot tarred boats are used for fishing during a season that extends from 1 March to 31 August. Boats are anchored to iron stakes embedded in the river bank and a fishing session can begin as soon as the stakes appear above the surface of the water, approximately two hours after high water. A rope, known as a warp, is attached to the stake and the boat can be taken across to the other side of the stream, where the other end of the warp is fixed to an anchor. The boat is secured along the line of the warp through a hook low down on its side and held at the bow by a raised stem piece and at the stern by a thole-pin. In operation, the boat and its net can be taken backwards or forwards along the warp if the wake of the salmon is observed. A strict rota is followed to determine fishing stations. The first team to arrive at the fishing station at Little Milford begins to stake No.1 – The Bite, while at the next tide it has to go to stake No.2 – The Lake, and so on. The stakes on the western Cleddau are: The Bite, First, Second and Third at Lake, The Quay, The Drot, Level Stake, The Stones, The Rail (where the stake is a piece of an old rail) and The Grimbank. On the eastern Cleddau, the stakes are Home Stake, Kerlisky, Layer's Park, Osmond, Edwards and Crafty.

The number of compass-netsmen in Dyfed has declined considerably for in 1939 there were 21 boats engaged in the trade and the fishermen

A compass-netsman (Glyn Morgan of Llangwm) fishing on the river Cleddau, 1972

combined compass-netting with seine-netting for salmon and drifting for herrings in the Haven. In 1860, a total of 70 compass nets were licensed in the Haven. During the early twentieth century many would also spend a part of the year sailing in deep-sea trawlers from Milford.

Seine-netting

The shore seine is widely used throughout Britain, particularly in river estuaries, and it is a simply constructed plain wall of netting 200 yards or more in length and of a depth suitable to the water in which it is used. It is important that the net should extend as far as possible from the surface of the water to the bottom and should stand as vertically as possible in the water. The head-rope is fitted with corks or plastic floats and the foot-rope weighted with lead or stones. The net is carefully stowed on the flat transom of a small boat. One of the crew stands ashore holding a rope attached to the end of the net. The boat is then rowed out into the stream from the shore on a semi-circular course, with the net being paid out over the stern. When all the net has been shot, the boat returns to the shore whence it set out. The crew then lands, the boat is made fast, and the net is hauled in. The landing place is usually downstream of the shoreman, but it is occasionally shot upstream, particularly if the boat, as on the Teifi and Dee, is equipped with an outboard motor. The hauling of the net is rapid and smooth and the two ends of the net are brought close together thus making a narrow bag of the middle of the enclosed space, where the fish are concentrated and can be hauled ashore. The foot-rope is hauled faster than the headline, thus making a more pronounced bag of the centre of the net. The whole net is drawn in; the salmon or sewin caught in its mesh are killed with the knocker and the net has to be re-arranged on the boat transom again ready for the next shot.

Despite its occurrence in estuarine waters throughout Wales, there are local variations in the dimensions and techniques of using the seine-net. Rowing boats equipped with outboard motors are used by the seine-net fishermen of the Teifi estuary. In the past, these boats were referred to by the fishermen as *Llestri sân* (seine vessels). Undoubtedly seine-net fishing has been well known on the Teifi for many centuries. George Owen, writing in 1603,[5] for example, describes the 'great store' of salmon 'as allso of sueings, mullettes and botchers, taken in the said Ryver neere St. dogmells in a sayne net after everye tyde'. Until the twentieth century salmon and sewin fishing in the spring and summer and herring fishing in the autumn winter seems to have been the main occupation of the inhabitants of St Dogmaels. 'It affords employment', said one nineteenth-century author[6] 'to such of the inhabitants that are not engaged in agricultural pursuits'. It is difficult to estimate the extent of seine-net fishing of the Teifi in the past, but in 1884 there were 62 licensed netsmen on the Teifi[7] and large quantities of salmon and sewin were taken. During the summer of 1883, for example, 'as much as half a ton of salmon was caught in the lower reaches of the river and sent to London and other places in one day'.[8]

During the course of the twentieth century, and particularly since 1939, there has been a steady decrease in the number of seine-net fishermen operating in the Teifi estuary. In 1939 there were 13 boats, each one manned by a team of 5 fishermen, operating on the Teifi; today the number has decreased to 2 boats, each one manned by 4 men only. In the 1920s, each team consisted of 7 fishermen and it has been estimated that 20 boats were engaged in seine-net fishing at the time. The 24 full-time fishermen

Seine-netting for salmon and sewin in the Teifi estuary in the 1930s
(Photo – M.L. Wight)

James Sallis of St Dogmaels hauling in a seine-net at Patch in the Teifi estuary

of St Dogmaels are all members of the St Dogmaels Seine-net Fishermen's Association, whose headquarters are at the Netpool in the village.

There are a number of reasons for the decline of seine-net fishing in the estuary of the Teifi. The fishermen themselves tend to blame the old River Authority for severely curtailing their activities. The season extends from 1 March to 31 August, but pre-1939 it commenced on 16 February. There are also limitations on the period of legal time for fishing, for no netsman may operate between 6 a.m. on Saturday and midday on Monday. High and ever-increasing licence fees are also blamed for the decline in the number of fishermen. Undoubtedly one of the main reasons for the decline has been the gradual silting up of the river estuary and a change in the course of the river. Silting has followed the decline of Cardigan as a sea port, for when the wharfs of Cardigan flourished a navigable channel was kept open to the sea. Sand from the estuary was also used as ballast for the ships. As a result of the decline of sea trade, the river has silted up and there would literally be no room for the 20 seine-nets of the past. Pools that yielded appreciable quantities of salmon have silted up and the seine-netting today is concentrated in four pools only – Pwll Nawpis, Pwll y Perch, Pwll y Castell and Pwll Sama.

To decide which fisherman goes to which stretch of river, lots are always drawn before each tide at the Netpool, by picking numbered stones (*cymryd shot*) from a hat. For that period of fishing a particular stretch of the river, known as a *bwrw*, belongs to the team that has drawn the stone with a number corresponding to that of the pool. No fisherman is allowed to fish any portion of the river other than the 'cast' or 'station' which fell to his lot before leaving the Netpool (*Pwll y Rhwyd*). The stretch of river surrounding the deep pools, where fishing is allowed, is known as *traill* (trawl) and the trawls of the estuary below Cardigan Bridge are known as Tram y Brain, now never used; Tram Fach below Pwll y Castell; Tram Pwll Nawpis, and Tram y Rhyd above Pwll y Perch.

Fishing takes place on the ebb tide and each team proceeds to its allotted station in the river. The team or teams that have drawn Pwll Sarna have a journey of over a mile to the bar of the river, while those that have drawn Pwll y Castell have little way to travel. The sailing down river from St Dogmaels is described as *mynd i'r san* (going to the seine).

The teams wait at their allotted station until the tide is considered right for the first cast or *ergyd*. With the net carefully folded on the stern, the boat, equipped with an outboard engine, is steered out to the centre of the pool, the net being paid out to a half-moon shape from the shore. The regulations state that 'One end of the head-rope of a draft or seine-net shall be shot or paid out from a boat, which shall start from such shore or bank and shall return thereto without pause or delay, and the net shall thereupon forthwith be drawn into and landed on the shore or bank on which the head-rope is being held'. One member of the team, therefore, acts as a shoreman and holds the head-rope on the bank; another is concerned with steering the boat while the other two payout the net from the stern of the boat. One of these is concerned with the head-rope and its cork floats, another with the bottom rope and its lead weights. Slowly the boat makes its way downstream and after the whole net has been extended, the boat is steered towards the shore. The boat is then anchored and the three fishermen together with the shoreman pull in the head-rope and net. If a salmon or sewin is caught in the net it is knocked on the head with a wooden *cnocer*. The net is then rearranged on the gunwale of the boat and is cast again. The process is continued until the tide makes it impossible to continue. A single ergyd takes approximately fifteen minutes. On returning to the village, the catch is weighed and packed ready for market. The fishermen are paid a piece-rate wage and the share of each team is divided into five equal parts – four for the fishermen and 'one for the boat'. The fifth share is designed to pay the expenses of the boat owner, who is, in most cases, a member of the fishing team.

A drift-net consists of a wall of netting shot from a boat across the current and allowed to drift freely. One end of the net is attached to a floating buoy or staff and the other remains fixed to the boat. The head-rope is corked and the foot-rope leaded, to keep the net upright.

The most important drift-net fishing area in the Bristol Channel is the Usk estuary, where eight boats, each crewed by two men, are licensed to use drift nets. In most cases the licences have been passed down from father to son over many generations; the principal family being the Sully family. The number of licences is limited to eight, and it is only when a fisherman dies or gives up work that the licence can be taken up by another. For the Usk estuary the boats drift upstream on the Severn with the flood tide as far as Portskewett and drift back again into the Usk on the ebb.

In the Dee estuary, trammel-nets each 100 yards long with a lint of 2¼-inch mesh and armouring of 11-inch mesh are used by four teams of fishermen. The trammel is a most difficult net to fix but, when in good working order, is said to be the most deadly of all nets. The 'Connah's Quay Trammel', as it has been called, consists of a central wall called the 'lint', of a loosely hung net, too small to gill the fish, and one or two outer walls of much larger mesh, called the 'armouring', which are not so slackly hung as the lint. All three walls are set to a common rope at the top and bottom, the lint naturally hanging slackly between the armourings. In the 1980s, the lint of a Connah's Quay net was made of carpet threads with a mesh of 1 inch from knot to knot. The square-meshed armouring had a mesh of 10½ inches and was made from fine cotton. Today, most trammel-nets are of nylon. The Bithell family of Flint have been trammel-netsmen at least since 1600 and every year from early March until late August they trammel for salmon between Flint and Greenfield. On every ebb tide they drift downstream in their stoutly built rowing boats and drift back again to Flint on the flood tide. Each boat is manned by two fishermen and the net when not in use is kept on a platform at the stern of the boat. Square-meshed armouring is preferred to the usual diamond mesh and this is usually woven by the fishermen themselves for attaching to the lint which is factory produced. The Bithell family are full-time fishermen who combine salmon fishing with shrimp trawling in winter. Should the salmon season be a poor one, the fishermen return to shrimping in the summer. A certain amount of trawling for flat fish beyond the Dee estuary is also undertaken by the fishermen.

A member of the Bithel family of Flint operating a traditional trammel-net in the Dee estuary

Salmon traps

Undoubtedly the trapping of fish is one of the oldest methods of fishing known to man, and throughout Wales, in rivers and along the sea-shore, may be seen the remains of stone or wattle weirs (*goredi*) erected many years ago for the capture of salmon and other fish. At Llanddewi Aber-arth on the Ceredigion coast, for example, a stone *gored*, 200 yards in length, operated until 1925.[9] In 1860 there were a dozen stone weirs operating between the mouth of the Aeron and the mouth of the Arth, while in the Menai Straits in north Wales, a number of weirs operated until recently. In addition to permanent, stoutly constructed weirs, all of which except for one on the Conwy have fallen out of use on Welsh rivers, there were, in the past, a large number of removable 'traps, hecks, crucks, cribs or inscale'. Salmon fishing was forbidden in many of these by the Salmon Fishery Acts of 1861 and 1865[10] except under grant or charter or by the right of 'immemorial usage'. By the Act of 1865, Special Commissioners were appointed to enquire into the legality of all 'fixed engines' used for catching salmon. 'That such engines were established by grant, charter or immemorial usage had to be proved to the satisfaction of the Commissioners who then issued a certificate of legality. In this way the number, size and position of all salmon weirs... became fixed for all time. No new weirs may be established nor may the position of the existing ones be altered.'[11]

The basket traps used in the Severn estuary at such places as Goldcliff and Porton are known as 'putchers', each putcher being a woven willow or wire basket, 5 feet to 6 feet in length and about 2 feet in diameter at the mouth, tapering to 5½ inches to 6 inches at the tip. Narrower baskets to fill in irregular gaps in the weir were also made. Each basket is open in weave and traditionally autumn-cut withies, usually from a six-acre withy plantation at Llanwern in Gwent, provided the raw material for the putcher maker. The baskets were made by the fishermen themselves during the close season, between 14 August and 1 May.

The traditional all-willow putcher was gradually superseded, firstly by galvanized wire baskets, later by baskets made from sea-resistant aluminium wire. Although galvanized wire and aluminium putchers had to a large extent replaced those of locally grown withy by the 1960s, traditional measurements are still adhered to by the manufacturers.

Willow putchers were still made and used by the fishermen at Beachley near Chepstow until 1990. To make a putcher, a low bench 18 inches high and 24 inches square is used. Nine holes are bored in the surface of this bench to form a circle 10 inches in diameter. Willow is split into three sections using an oak cleaver, held in the palm of the maker's hand. The nine rods required are passed through the holes in the bench, and a withy ring, approximately 10 inches in diameter, is plaited around the rods, close to the surface of the bench. Nine shorter rods, either cleft or in the round, are then inserted into the ring and another ring is plaited about half way up the rods and another near the top. The ring for the nose is plaited and coiled back in between the rods to be attached to the lowest plaited ring on the bench. The putcher is removed and the base ring, approximately 24 inches in diameter, is woven. The putcher is then ready for use. Withy baskets were expected to last for two fishing seasons and possibly a third season after repair.

The method of using putchers entails the construction of a stout timber framework built across the main tidal flow of a river. The framework carries row upon row of the cone-shaped putchers in which fish are trapped and then stranded on the falling tide. The high rise and fall of the tide in the Severn estuary is an important factor in allowing a profitable number of baskets to be submerged along ranks of workable length.

A putcher rank in the Severn estuary at low tide

A method of fishing that disappeared in recent years from the rivers of south-east Wales is that of using single rows of basket traps known as 'putts'. Putt weirs may still be seen at such places as Oldbury in Gloucestershire and a weir of this type existed at Goldcliff in the 1930s, when it was used for catching a variety of fish ranging from salmon to shrimps. Indeed, according to the late Wyndham Howells, a Goldcliff putcher-fisherman, the main function of the putt weir was to catch shrimps that were actually boiled at the fishery, but large quantities of flat fish were also caught in the weir.

Unlike the putcher with the open weave, the putt is a closely woven basket trap, consisting of three sections: 'kype', 'butt', and 'forewheel'. It is made of willow, hazel or whitethorn. A single row of perhaps 120 putts is used in this type of weir. The kype may measure between 5 and 6 feet in diameter at the mouth, and the whole trap from 12 to 14 feet in length. The three sections of the putt are kept firmly in place on the river bed by a series of wooden stakes with a forked stick driven firmly into the ground keeping the forewheel in position. The catch may be removed by taking out a wooden bung at the front of the forewheel and emptying the contents into a shoulder basket, known as a 'welsh' or 'witcher'. The mouth of the trumpet-like kype faces the ebb tide and is usually located at the tail of a pool in the river.

A putt rank at low water in the Severn estuary at Shepperdine (Photo – City of Gloucester Museum)

The entry or 'kipe' of a three-sectioned putt

3 HERRING FISHING

The herring (*Clupea harengus*) has always been of considerable importance in the history of coastal communities in Wales. Not only was it a main constituent of the diet of the people, providing cheap, nourishing food, but it was also an important item of export. Indeed barrels of salted herrings were about the only items that were available for export from many a coastal village. The pursuit of the herring was for many centuries central to Welsh fishing activities.

The herring is a pelagic fish that occurred widely along the coast of Wales, especially Cardigan Bay and the northern coast. Two coastal villages in particular – Nefyn in Gwynedd and Aber-porth in Dyfed – became particularly well-known as centres of herring fishing and *Penwaig Nefyn* and *Sgadan Aber-porth* were famous throughout the country. Today, hardly anyone in those villages is concerned with herring fishing and the days when deep-sea, sailors made a point of being home in the autumn of every year in order to take a part in the herring fishing from their native villages have long passed. Along the whole of the Welsh coast, there has been a sharp decline in herring fishing within the last forty years or so, and it is said that the vast shoals of fish that made an appearance along the Welsh coast no longer come. Some of those shoals according to tradition were as much as seven miles in length: in the seventeenth century the coast of Pembrokeshire was said to be 'inclosed in with a hedge of herrings'.[1] There was no creek or bay without its fleet of herring boats. In recent years, the herring, often regarded as the food of a poverty-stricken community, has become far less popular in the diet of Welsh people.

Although the decline in herring fishing from the coastal villages of Wales was apparent from 1914, considerable quantities of herring were still being landed in Milford Haven in the 1920s. In 1925, Milford became, for the first time, the principal landing station for herrings in England and Wales. At that time and throughout the 1930s, a large proportion of the catch was landed by Scottish and east-coast drifters that visited the Irish Sea in the autumn and winter months in pursuit of the herring.[2] Many of these drifters came to the Irish Sea as an alternative to fishing the depleted herring grounds of East Anglia after the end of the Scottish herring season that extended from June to September.

Since 1945, there has been a steady decline in the number of herrings landed at Welsh ports and since 1972, the herring has hardly figured in statistics relating to landings in Milford. Between 1949 and 1959, herrings were landed at Cardiff and Swansea as well as Milford and a substantial proportion of the landings were made by deep-sea trawlers as a by-catch of demersal fishing at the Smalls, off the southern and north-western coasts of Ireland and the Minches. With the collapse of hake fishing in the mid-1950s, the landing of herrings at Cardiff and Swansea ceased completely, but for a few years, from 1955 to 1958, landings at Milford increased as a result of Lowestoft drifters being based at Milford to pursue fishing in the Irish Sea. Both the trawl-net and drift-net fisheries at Milford Haven were effectively killed off in 1960 by the unilateral introduction of baselines by Eire across, among others, Dungarven Bay, which enclosed the main inshore herring grounds. All other herring landings in Wales during the post-war period were made by inshore vessels which have been a mixture of northern Irish, southern Irish, Welsh and occasional Scottish vessels. In the 1960s a substantial proportion of herrings landed at Milford were by Irish trawlers and those at Holyhead were entirely by northern Irish vessels; mainly from the fishing port of Kilkeel.[3] After 1966 however, landings at Holyhead virtually ceased. The main port for herring landings within the last ten years has been Conwy and trawlers actually based on that port have been responsible for landing herrings mainly as a by-catch of the winter whiting fishing. They were also engaged in the summer herring fishing season off the Isle of Man coast and landed much of their catch in the Isle of Man. All other landings at the smaller Welsh ports or harbours were made by part-time fishermen using gill-nets drifted from small boats or set across harbours.

Life history of the herring

The herring differs from other sea fish of commercial importance in that its spawn is demersal; that is, it is deposited on the sea bottom. The fish itself is pelagic; that is, it swims near the surface of the water. Most herrings were caught in drift-nets, but since the fish occasionally occurs in much deeper water, trawl-nets were sometimes used for their capture.

In the past, it was believed that the herring migrated from Arctic waters. 'The great winter rendezvous of the herring is within the Arctic Circle' said Pennant.[4] However, this theory has been largely disproved and it is now believed that the herring reaches maturity and spawns close to its birthplace.

Writing in 1796, Pennant claimed:

... This mighty army begins to put itself in motion in the spring ... They begin to appear off the Shetland Isles in April and May; these are only forerunners of the grand shoal which comes in June. The first check this army meets in its march southwards is from the Shetland Isles, which divide it into two parts; one wing takes to the east, the other to the western shores of Great Britain. Those which take to the west ... proceed towards the north of Ireland where they meet with a second interruption and are obliged to make a second division; the one takes to the western side and is scarce perceived being soon lost in the immensity of the Atlantic; but the other which passes into the Irish Sea, rejoices and feeds the inhabitants of most of the coast that border on it.

Although few believe that shoals of herrings migrate at certain times of the year, it is nevertheless a fact that there have been periods when the herring has almost been completely absent from the coast of Wales. In Pennant's day, they were 'almost capricious and do not show an invariable attachment to their haunts. We have had in our time' he adds, 'instances of their entirely quitting the coast of Cardiganshire and visiting those of Caernarfonshire and Flintshire, where they continue for a few years, but in the present year they have quite deserted our sea and returned to their old seats.'

There are two periods of spawning; one in the spring and the other in the autumn. The herring that spawn in the spring, spawn near the coast in water of low salinity and they may even spawn in fresh water. The spring herring is still caught in the upper reaches of Milford Haven by fishermen from the villages of Hook and Llangwm. Using the small, tarred Llangwm rowing boats and fine meshed, weighted, gill-nets each 150 feet long and 20 feet deep, they catch substantial quantities of herrings. The fishing season lasts for about a month in late February, March or early April, depending on the temperature of the water. 'It is only now we are beginning to realise', said Professor E.W. Knight-Jones[5] 'that this comparatively small body of water, may nevertheless be of great importance in the production of herring throughout the southern Irish Sea, because they are so concentrated there.'

Autumn spawning takes place well away from the coast in deep salt water and it was the capture of the autumn-spawning herring that was the main pursuit of the Cardigan Bay and north Wales fishermen. Fishing took place between August and February, although local lore and taboos may have limited fishing to certain periods within that season. At Aber-porth, for example, fishing could not begin until thousands of sea-birds, known locally as *Guto Gruglwyd* made an appearance on the coast and began congregating above the herring shoals. This was usually in September.

Fishing had to cease by Christmas, although some fishermen believed that the second week in November should mark the end of the herring fishing. 'Mae "'sgadan Glangea" yn aml wedi torri bolie' (The November herring has often spawned) said one informant.[6] At Aberystwyth on the other hand, the second week in November, particularly if an east wind was blowing, was the best time to fish.[7] At Moelfre, in Anglesey, the herring season extended from October, after the harvest and thanksgiving services, to February,[8] while at Nefyn[9] September to January was the recognized fishing season, and if a north wind blew, then the fishing was said to have been good.

The preservation of the catch

Usually, herrings are scattered in deep water, but at certain times of the year, they come together in vast shoals, close to the coast, for the purpose of spawning. They can then be caught in drift nets, but once they are caught, the fish are very perishable and have either to be sold and consumed very quickly or they have to be preserved by some means or other.

In Wales, three categories of herring are traditionally recognized – Fresh, Bloatered and Red.

1. *Fresh Herrings*

In coastal communities, fresh herrings were a favourite food but it was considered important that the eyes of the herring were red and bright, which denoted freshness. If the eyes were dull and grey, then the fish was stale and could not be eaten. A fish that had already spawned was not in demand and many fishermen threw these herrings back to the sea or used them as a bait in long lining. In Llŷn, spawned herrings were known as *soldiars* and there was no demand for them, for according to the salesmen of Nefyn herrings, the characteristics of a good fish were:

'Penwaig Nefyn, penwaig Nefyn
Bolia fel tafarnwyr
Cefna' fel ffarmwrs'

(Nefyn herrings, Nefyn herrings
Bellies like inn keepers
Backs like farmers)

At Aber-porth, fresh herrings were sold with the cry

'Sgadan Aber-porth, sgadan Aber-porth
Dau fola [llygad] ac un corff'

(Aber-porth herrings, Aber-porth herrings
Two bellies [or eyes] and one body)

Fishermen usually sold the major part of the catch to fish-merchants. In Llŷn, these merchants were known as croeswrs and in the 1920s, when the herring industry flourished, they purchased herrings from the fishermen at an average price of 5s. per half a hundredweight and sold them to farmers at an average price of 2d. per fish. Individual farmers either salted or kippered the fish as soon as possible after they had purchased them from the croeswrs. At Amlwch, the fish-merchant was known as a *tsiecmon* and most of them came from Bangor and Llanerchymedd. By 1920, most herrings were sent by train from Amlwch and Benllech stations directly to fish-merchants in Liverpool.

At Aber-porth, there was a great deal of direct buying from the fishermen: farmers and others came to the beach with carts to purchase herrings which they salted at home. Fish-merchants, known locally as carriers, also purchased fish for selling in the countryside. Each carrier would buy herrings from a particular boat and if a particular boat crew had contracted to supply a carrier with fish, then that carrier could not approach another boat crew for supplies. Herrings were purchased by the long hundred (*cant fowr*) which was 120 fish or by the meise (*mwys*) which consisted of five long hundreds.

2. *Bloatered herrings*

Fresh herrings can be lightly pickled in brine or dry salt as soon as possible after capture and occasionally they are lightly smoked to preserve them for slightly longer. The traditional method of bloatering at Aber-porth was to place the fish on the flagged floor of one of the two salting houses near the beach. A quantity of dry salt was sprinkled over the fish and the herrings had to be turned at frequent intervals until they had absorbed a considerable amount of salt. When the fish became supple enough so that the head of each fish could be bent to touch the tail, the herrings were ready for the next stage.

The salting house was cleaned out and the fish re-arranged very carefully on the floor in layers with salt in between and left for three weeks. They were then washed in clean water and again returned to the salt house with dry salt placed over them, for a few days. If desired, the bloaters could be steeped in a vessel containing clean water for three days and three nights, with a daily change of water. They were then placed in the open air to dry thoroughly.

On the Llŷn peninsula, although there were salf houses in many coastal villages, bloatering was usually carried out in individual houses or farmsteads, using a wooden cask or other vessel for the purpose. Quite often, brine rather than dry salt was used and the brine (*y bicil*) was made by mixing salt and water. In order to test the salinity of the brine, a potato was dropped into the mixture; if it floated, then the brine was strong enough for preserving herrings.

The usual method of salting was to place the herrings in a cask with layers of dry salt in between. The fish were turned at frequent intervals until they had absorbed the salt. Each fish was then wiped with a piece of cloth and replaced in the cask with salt, but not turned in any way, to remain there for nine days. It was a common practice in the Nefyn district to dry the bloaters – *penwaig sychion*. A thick layer of ferns was spread on the ground on a sunny day and the salted herrings were then packed in willow hampers, with coarse salt, ready for sale. Each hamper contained 100 to 150 fish and quite often *penwaig sychion* were taken by ships to Caernarfon and other places.

3. Red herrings

Although there were modern smokeries at Milford Haven until 1990 the traditional method of producing red herrings (*sgadan coch* or *penwaig cochion*) was somewhat different. In kippering, fresh herrings are split and cleaned, pickled lightly, and smoked in a specially constructed smoke house, preferably with smoke from oak chips. In making red herrings, however, the fish are not split and the preliminary salting may last for five or six days before washing and hanging in a smoke house for another five or six days. In some areas, notably the Nefyn district, it was customary to give herrings a second smoking for longer preservation. After the first smoking, the herrings were again placed in brine and then smoked for a further five days.

In many parts of the country, smoke houses were commonplace. At Nefyn, in the early nineteenth century, for example,[10] there were numerous curing houses on the foreshore, and throughout the centuries 'red herrings' were important items of export from the village. In 1685, for example, the Thomas and Jane of Porthdinllaen took 80 barrels of salted herrings and 18 barrels of red herrings to Chester, and in the seventeenth and eighteenth centuries important items of import into Nefyn were 'empty barrels for the fishing industry'.[11] At

Llandegai, the curing house built in the early nineteenth century, 'has proved almost instantaneously a great and extensive advantage to the poor, who are now enabled to buy herrings at a reasonable rate'.[12] At Criccieth, herrings were salted and smoked, 'but with no great attention or skill'.[13]

At Aber-porth, smoking was often carried out in the open chimneys of individual houses. Salted herrings were hung through the eyes on wooden sticks or wires and placed in the chimney to smoke for at least two hours. Oak leaves and ferns were greatly favoured as fuel.

At Milford, a smoke house was built in 1904 to kipper herrings brought in by the 70 drifters that used the port. By 1922, the herring fleet was increased to over a hundred and a second smoke house was built in 1923 and another in 1924. Two more were built in 1925. Protests by some of the inhabitants against the alleged nuisance caused by these additional smoke houses called forth a leaflet from the representatives of the trawler owners and fishermen. It asserted that any attempt to hamper the development of the herring trade might have a serious effect on those whose living depended on the fishing industry. The herring trade, it was pointed out, was growing at Milford, and if encouraged, the town might become one of the greatest centres of it in the country. A conference was called on 6 May 1925 to discuss the matter, but without any result. Three years later, an action for damages was taken against three firms which operated smoke houses. It was alleged that they caused to issue from their respective premises 'offensive, poisonous and unwholesome smoke vapour'. The hearing at the Carmarthen Assizes lasted for three days and a large number of witnesses were examined. The upshot was that the judge found that the smoke did not constitute a nuisance in the legal sense. Those who elected to live in a fish town must expect to experience an inconvenience of that kind.[14]

The history of herring fishing

Undoubtedly, herring fishing in Wales is of considerable antiquity especially in Cardigan Bay. In 1206, it was said in *Brut y Tywysogion*,[15] that great quantities of herring were landed at Aberystwyth – 'y roddes Duw amylder o byscawt yn Aberystwyth ac nabu y kyfryw kynno hynny' (God provided more fish at Aberystwyth than ever before). More often that not, herring fishing in Cardigan Bay was carried out by part-time fishermen, part-time farmers who operated in small rowing or sailing boats. Pastoralism seems to have been the main occupation of the inhabitants of the Welsh shores but they were also engaged in herring fishing in late summer and autumn.

An Aberystwyth three-masted herring boat of a type widely used in the port until the 1880s (Photo – R.J.H. Lloyd)

Sea fishing appears to have been, in the main, free in medieval Wales, but at Aberystwyth a meise of herrings, the 'prisemes', was demanded of every herring boat landing. About the middle of the fourteenth century, more than twenty fishing boats were employed at Aberystwyth. At this time, Beaumaris, Barmouth and Aberdaron were the principal herring ports in North Wales and Tenby, Pembroke and Haverfordwest in the south. A considerable quantity of fish, white and red herrings, hake and salmon, would appear to have been imported from Ireland by Welsh trading vessels.[16]

Aberystwyth was almost certainly the principal herring port of Wales in the Middle Ages and herring fishing developed into an important commercial venture, with substantial exports of salted herrings to Ireland and elsewhere. The fishermen of Aberystwyth were expected to hand over a proportion of their catch of herrings to the Lord of the Manor and an extant Court Roll of the Aberystwyth Borough gives an interesting glimpse of fishing conditions. It seems that,

> … in one year there were between twenty and thirty cases connected with the fishery. Some of the delinquents persisted in selling their herrings on the sands below high water mark in order to escape paying market tolls; others would seem to have taken part in the herring industry without obtaining properly accredited licences for their fishing boats. Heavy fines were sometimes imposed and led to heated altercations between the mayor and the fishermen.[17]

Inshore fishing, particularly in Cardigan Bay, developed phenomenally in the sixteenth century and every creek and bay had its fleet of herring boats. The requisites for fishing; the nets for catching and salt for preserving were imported from Ireland in Irish-owned boats, that for centuries had been engaged in trading along the Welsh coast.[18] Casks for preserving and exporting salted herrings were produced in most coastal neighbourhoods by the large number of coopers who lived in all parts of the county. An indication of the growing importance of herring fishing is given by George Owen in his Description of Pembrokeshire of 1603.[19] He describes the vast shoals of herring around the coast of his county 'which being in great store and sold to parts beyond the sea, procureth also some store of money.' He adds:

> This fishing is chiefly from August till neere Christmas, the middle of first fishinge is counted best as that which is fullest and fattest. The order of taking them is with drovers [that is with nets which drift with the tide] and shootteings of nettes in known places chosen especially for the fairness of the ground, which nettes are shoote in the evening, the later the better, and drawn up in the morninge with such store of fishe as pleased God to send.

Large quantities of herrings were sold at Haverfordwest, Pembroke and Tenby markets and 'the fish at Tenby, where there was a daily market for it was held in special estimation'.[20] The Welsh name Dinbych-y-pysgod (Tenby of the fish) is in itself indicative of the importance of fishing in the local economy.

Herrings, throughout the centuries, have been sold by 'the meise' (*mwys*) which consisted of five 'hundred' herrings. The 'hundred' consisted of 120 herrings and the usual method of counting was to count two score of fish, throwing one fish aside as a tally. After counting five score of herrings another herring was thrown aside to denote a hundred. A meise, therefore, consisted of 120 x 5 herrings = 600 + 15 'warp' (the herring thrown aside to keep tally of the number of hundreds). The total in a meise was therefore 620 herrings. Usually, the herrings were counted in threes – a *bwrw* or *mwrw*. Two herrings were taken in one hand and one in the other and this was repeated forty times to produce a hundred.

During the seventeenth and eighteenth centuries, the inshore herring fisheries of Wales developed substantially after the enactment of the Bounty Act of 1705 and the subsequent Act of 1718. Bounties of 'a shilling were paid on every barrel of shotten red herrings exported beyond the sea; one shilling and nine pence of every barrel of full red herring and two shillings and eight pence on every barrel of white herring'.[21] The Act for the Encouragement of the British White Herring Fishery of 1750[22] resulted in a phenomenal increase in inshore fishing. By this Act, a bounty of 30s. per ton was paid for every decked vessel of 30 tons to 80 tons. There was a rapid increase in the number of decked sailing vessels operating from Welsh ports. Another Bounty Act of 1787 redressed the balance by granting a bounty on every barrel of herrings taken by open or half-decked boats of less than 20 tons. As a result of government action the herring fishing industry developed very rapidly during the second half of the eighteenth century. In the Llŷn peninsula, for example, Pennant speaks of the abundance of herrings taken 'from Porth Ysgadan or the Port of Herrings to Bardseye Island'.[23] He expressed the opinion that agriculture in Llŷn suffered very greatly from the concern of the people with fishing, but writing a few years later,[24] 'was satisfactorily assured that no injury could accrue ... in as much as the harvest was always either housed or at least saved before the coming of the fish'. Many of the fishermen of Wales were, of course, part-time farmers, part-time fishermen.

After 1790, Swansea developed rapidly as a fishing port and market, but there had been a decline in the fortunes of Aberystwyth as a herring port. Earlier in the century, the town still reigned supreme as the most important of herring ports. One eighteenth-century writer noted that the main contribution to Aberystwyth's wealth was its fishing trade of Cod, Whitings but principally Herrings... the Herring Fishery here is in most so exceedingly abundant that a thousand barrels have been taken in one night.'[25]

By the end of the century, the prosperity of Aberystwyth had waned and the herring that occurred so profusely in Cardigan Bay was in 1744 'a stranger to the coast'.[26]

Tenby, another important fishing port in earlier times, had declined. 'The fishermen were incapacitated by poverty, their property consisting only of a few open boats with nets and dredges ... quite unsuitable for venturing to sea and capable only of working in the bay or just outside the pier.'[27] Tenby waters were however visited by Brixham and Torbay vessels who transported their catch to Bristol market.

Although Swansea, and later Milford, developed as important fishing ports in the nineteenth century, much of the fishing in the Cardigan Bay area was carried out by part-time fishermen operating from open boats. In Cardiganshire, for example, coastal villages such as Aberaeron, New Quay, Llangrannog and Aber-porth supported many fishermen. In the 1830s Aberaeron had 'a lucrative herring fishery in which about thirty boats with seven men to each are engaged'.[28] New Quay had fish 'of very superior quality' and the Teifi estuary had its fleet of herring boats with the village of St Dogmael's regarded

... as one of the principal stations for the herring fishery ... where the boats engaged in it are commonly from eight to twenty burthen with masts and sails, but mostly open, without decks and manned by six or eight men. The herrings generally make their first appearance on the neighbouring coast between the middle and end of September, which is considered the best period of the season, as they will bear carriage to distant markets and the harvest being commonly over, the fishermen can be better spared from agricultural labours.[29]

Herring fishing was certainly of great importance in Wales throughout the nineteenth and early twentieth centuries, although there were extended periods when few fish were caught. At other times there was a glut; to such an extent that at Nefyn in one year, herrings were used as a fertilizer for the land. When a period of scarcity came again, local fishermen attributed this to divine punishment for the misuse of food in a period of prosperity.

An essay written for the National Eisteddfod at Liverpool in 1884[30] gives an indication of the importance of herring fishing along the Welsh coast during the last quarter of the nineteenth century. In 1883 'vast shoals of herring appeared off the Caernarfonshire coast from Conwy to Bangor and most opportunely, for the stone (paving sett) quarries being short of work through the rapid introduction of wood pavement into the larger towns – the workmen thus thrown out of work, found employment in the herring fishery and were very successful.' Caernarfon was an important fishing port with 300 boats operating from the harbour. Nefyn:

... the quaint, old fishing and seafaring town ... is famous for the quantity of its herrings, this fish being found here in much better condition and of finer flavour than on the other side of the promontory of Llŷn – from Pwllheli to Criccieth. It is well known locally that the herrings taken along the northern coast of Llŷn are greatly esteemed arid fetch a good price. Those on the other side, in Cardigan Bay, are not much more that half their size. The one sort is known as Nefyn and the other as Criccieth herrings.[31]

In the northern part of Cardigan Bay, between Pwllheli and Aberystwyth, there had been a decline in herring fishing and Davies believed that 'the fish are farther out to sea whither these poor people with their little boats are not able to follow them'.[32] Aberystwyth had recovered slightly as a herring port, for after many years when very few herring shoals made an appearance on the coast, they had by 1883 become abundant again. Fishguard, Dinas, Cardigan and St Bride's Bay were important fishing stations and Milford

Haven was gradually developing to become by 1910 one of the principal fishing ports of Britain. The southern coast of Wales was not noted for its herrings although some were landed at Tenby and Swansea.

Despite the fact that substantial quantities of herring were still being landed in Wales in the late nineteenth century, the industry according to some observers was grossly neglected. 'From the extent of its seaboard of over 600 miles... one would imagine that this would be a source of profit, but as an industry it seems spasmodic and local. In north Wales a few fishing smacks put out for the mackerel and herring when they are supposed to be in the bay. Local markets are supplied, but no trade exists with the interior.'[33]

With the notable exception of Milford, herring fishing declined very greatly in Wales during the first quarter of the twentieth century. By 1914, both Nefyn and Aber-porth had ceased to be herring ports of significance, and it seems that the herring was once again a stranger to the Welsh coast. Milford drifters were able to fish in much deeper water than the small rowing boats that were commonplace in seaside villages, but in addition fewer men were prepared to follow what was, of necessity, a hazardous occupation. In most herring ports the crews of many fishing boats consisted of merchant navy men prepared to spend the autumn and winter months in their home villages as fishermen. 'Nothing can be done without men', says the Report of the Fisheries Superintendent,[34] 'the type of men who have usually carried out inshore fishing are gradually dying out, and are not being replaced by younger men'. The 1914-18 war hastened the decline of the Cardigan Bay fisheries and by 1929 'the Cardigan Bay inshore fisheries are in a very parlous position.'[35] At Pwllheli, 11 fishing boats were operating in the 1930s, as compared with 24 in 1914, but at Aberystwyth in 1928, there was only one first-class fishing boat. 'This boat of the usual Cardigan Bay type i.e. thirty-five to forty feet long and eight to twelve feet beam – sailing boats but with marine engine fitted within the last ten years or so.'[36] This boat usually operated between Aberystwyth and New Quay, but never a greater distance than twelve miles from the coast. In addition, about 9 small rowing boats were concerned with fishing close inshore between Borth and Aberaeron during the herring season.

Holyhead was also prominent in the 1920s as a herring port, but most catches were landed by Scottish drifters that fished the Irish Sea.

By 1939, herring fishing had ceased to be an major occupation along the Welsh coast. On the east coast of Anglesey, serious herring fishing stopped in 1939 due to the almost complete absence of fish. One fisherman spent ten days fishing and as a result of his efforts he caught one fish only.[37] New Quay in Dyfed, to quote another example, was an important fishing port before 1914 and in 1913 a total of 938 cwt. of wet fish were landed in the harbour. By 1937, this had declined to 44 cwt. Caernarfon in 1913 landed 1,842 cwt. but by 1937 this had declined to 25 cwt.[38]

Since 1945, there has been a steady decline in the herring catch, except that both Milford and Holyhead, and to a lesser extent, Bangor, enjoyed a golden era in the late 1950s and '60s as a result of the efforts of the Irish and Scottish fishermen. But in Wales as a whole, the herring has ceased to be of importance.

The following is a random sample of herring landings in Welsh ports between 1946 and 1973:[39]

Herring catches (in hundredweights) at Welsh ports

	1946	1949	1952	1958	1963	1967	1970	1973
Milford		1,416	16,920	23,922	45,656	7,874	9,962	4,828
Swansea		1,217	1,217					
Cardiff	1,416	2,657	513					
New Quay	40	8	72	1				85
Aberystwyth								51
Holyhead	65	27		20,869	6,203			
Conwy				290	791	35	701	524
Aberdyfi								67
Pwllheli			1		9			
Bangor							703	1

Boats

Most of the larger boats used for herring drifting in Wales were basically of the 'nobby' type of decked vessels, regarded as boats of excellent sailing quality. Although many of these were built in Welsh boatyards, a number were brought to Welsh ports from Scotland and Lancashire. Most of the boats were carvel-built with cutaway stems and steeply raking stern posts. The boats were built with little sheer and had the appearance of sitting low in the water. The counter was sometimes square, but on later boats it was rounded. The vessels were usually cutter rigged, but sometimes boats were converted to ketch rigging. The cutter rig usually consisted of mainsail, foresail and jib, but generally there was no topsail. Herring drifters varied from about 25 feet to 40 feet in length, although vessels of 50 feet were commonplace.

The method of fishing with these drifters was to set out for the fishing grounds in the evening, so that the nets could be set before dark. Drift nets floated like a wall in the water, the upper edge being kept up by corks or floats and the bottom weighted with lead. In the days of sail, the boat was put before the wind and the nets shot over the side. When they were all out, the warp to which the nets were fastened was brought over the bow of the boat like a mooring and the vessel rode to it, keeping a pull on the nets which kept them stretched out in a line. While the boat was fishing, the sail was lowered and the head of the vessel was kept to the wind. Often, the mainmast was lowered to lessen the area of wind resistance.

There were, of course, many local variations in decked vessels in Wales. At Tenby, for example, many of the larger vessels belonged to Milford but were manned by Tenby men.[40] Most were cutter rigged and the smaller boats were of 16 or 17 tons, with a few of 30 tons or more. The smaller vessels were usually manned by two men and a boy, the larger by three men and a boy.

Luggers, open boats 15 feet to 25 feet in length, were also widely used for herring fishing from Tenby:

> Some boats had carvel built hulls, but most of the early ones were clinker built. They had straight sterns, long straight keels and slightly raking transoms. The forepart was decked to the mainmast and underneath was a small cuddy. There were usually three rowing thwarts and a thwart and quarter benches in the stern sheets. All the ballast was carried inside and they drew about 3 feet of water ... The mainmast was stepped about one third of the boats' length from the stern post and was clamped to a stout thwartships beam called the cross pillar, at the after end of the fore-deck. The mizzenmast was usually about half the length of the mainmast and was stepped in a socket on the transom, to starboard of the rudder head. It followed the rake of the transom. When a bowsprit was carried, it was a long spar which passed through an iron ring to starboard of the stern post with the heel secured in a socket on the fore-deck near the mainmast.[41]

Luggers of somewhat similar design were in common use in Cardigan Bay. Aber-porth herring boats, of which no single example has survived, were 25 feet or 30 feet long and although the boats were equipped with mainmast and mizzenmast, sails were never used for herring drifting. The boats did not possess rowlocks and five to eight men were required to man a single boat and each boat was equipped with about twenty nets. These heavy clumsy clinker-built boats had virtually disappeared from Aber-porth beach by 1905 and they were replaced mainly by Aberystwyth-built beach boats.

All along the Cardigan Bay coast, by far the most popular herring boat used during the twentieth century was a small double-ended beach boat produced by David Williams, the Aberystwyth boat-builder. They were clinker-built boats 18 feet long and 6 feet beam. 'They were primarily rowing boats for carrying passengers but they used a standing lug sail off the wind and engaged in herring and line fishing in their season. They were rather light craft, but were highly thought of by the Aberystwyth men who would venture a considerable distance in them.'[42] For herring fishing they usually carried a crew of two and each boat worked from six to eight nets. The first of these boats was produced in the 1890s and they soon replaced the older heavier boats in many of the Cardigan Bay harbours.

In addition to the double-ended beach boats that became so common in Cardigan Bay from the beginning of the twentieth century, Lloyd recognizes three other types that operated in the Bay.[43] The first used during the first half of the nineteenth century was a heavy clinker-built boat 25 feet to 28 feet long with bluff bows, full beam and a transom stern. 'It was an open boat, apart from a small fore-deck and had no thwarts, but a heavy transverse beam was placed a few feet aft of the midship section, dividing the main working space into two parts of about equal size. Like the Tenby boats, it was cutter rigged with jib, stays'l, mains'l and probably a lug-headed tops'l and had a very long bowsprit and a boom that extended well beyond the transom. The mast was stepped against the aft end of the fore-deck and was supported by two or three shrouds on each side and a foe stay.'[44]

The second type described by Lloyd as operating from Aberystwyth was a three-masted boat, possibly introduced into the area by Borth fishermen towards the middle of the nineteenth century. 'The clinker built hulls were entirely undecked and their long straight keels varied in length between 23 and 25 feet. The stern was fairly straight and the transom upright ... they were invariably sailed, though they did not sail well to windward and had to be helped round with an oar when going about.' The boats were equipped with three masts, but when herring fishing, the mainmast was removed and the boat sailed under foresail and mizzen only. The herring gigs were smaller versions of the three masters and they varied in length from 16 to 21 feet. Each boat carried a crew of two or three and they were widely used for taking holiday makers for trips around the bay.

In north Wales, a variety of sailing and rowing boats was used by the fishermen and most were built locally. At Nefyn, for example, a double-ended beach boat by Matthews of Menai Bridge was popular. This was 18 feet long with a 7-foot beam, while at Moelfre, boats of 20 feet to 30 feet, each manned by a crew of four were the most common. In 1914 a Nefyn boat could be built for as little as 25s per foot so that a boat would cost about £15.

In Llŷn, a wide range of boats was used for herring fishing. At Abersoch, for example, with its sheltered harbour, boats of the 'nobby' type, 32 feet to 34 feet long could be used, but at places such as Aberdaron, with little shelter, smaller boats had to be used. There were boats of excellent sailing qualities and although many carpenters in Llŷn were capable of building boats, the most notable was John Thomas, who had lived on Bardsey but who for many years until his death in the early 1960s, practised his craft at Aberdaron and Rhiw. The boats were usually 12 feet to 16 feet in length with a beam of 4 feet 6 inches to 4 feet 9 inches, and they were built specifically to sail in very heavy seas. They were equipped with rounded sterns and were very deep. Double-ended boats were never popular in Llŷn. Some of the boats, described as 'female boats' (*cychod banw*) were equipped with centre-boards and carried mainsail and jib with four oars. The male boat (*cwch gwrw*) drew far less water and was a rowing rather than a sailing boat, and the four oars were square and heavy. In a 14-foot boat, which seemed to be the size favoured by Aberdaron fishermen, the oars had to be at least 14 foot long. All the boats were built to a traditional pattern and the timber preferred by the builder was larch, preferably larch from an upland forest that was regarded as being much drier that that from lower forests. John Thomas himself was responsible not only for shaping all the planks by hand, but also for the actual felling of the larch trees.

The herring ports

Nefyn

Herring fishing on the north coast of Gwynedd is of great antiquity and the village of Nefyn, in particular, was renowned as a centre of the trade. An early nineteenth-century observer commented:

> Along the bay are built the curing houses for the herrings which are captured during the season in considerable numbers. Here also under the Point is a small pier for the protection of the fishing boats, of which about forty belong to this port. Each of these is commonly the property of seven persons who out of the season are either agriculturalists or tradesmen, or engaged in the coasters which supply the Liverpool market from the neighbourhood with poultry, shell fish or other smaller articles. At the commencement however, of

the fishery all other avocations cease and every thought and wish is directed to
the improvement of the advantages held out by these bounties of providence.

Long before the nineteenth century, Nefyn was an important fishing port.
As early as 1287 it possessed 63 fishing nets and herring fishing was widely
practised as a supplementary occupation to farming. Local inventories for
the year 1680[46] show the following entries:

Einion ap Addah – 9 oxen, 6 cows, 20 sheep, 3 heifers, 3 fishing nets.
Ieuan ap Madoc – 4 oxen, cow, horse, heifer, boat and 4 nets.
Llywarth Crun – 1 cow, 1 net.
Bleddyn Fychan – 6 oxen, 3 cows, 2 horses, 1 small boat, 3 nets.
Tagwynstl wraig Addaf – 2 cows, horse, heifer (3 years old), 1 net.
Dai Bach – 2 sheep, heifer, 2 nets.

In 1635, the major proportion of Nefyn's population of 60 men were
employed as fishermen and in 1771 the value of the annual catch at Nefyn
was £4,000.

By the end of the nineteenth century, herring fishing was fundamental to
the economy of Nefyn, and by 1910 it was estimated that there were about
40 boats, each manned by a crew of three or four operating from Nefyn. In
a four-manned boat, the 18-foot craft was rowed by three men with a pair
of oars each, while the other, usually the owner of the boat, was responsible
for the steerage. Each crew member had two or three nets. Although Hall
said that many of the fishermen from Nefyn sailed as far as the Irish coast in
search of herrings, undoubtedly the Gwynedd coast, up to three or four miles
from Nefyn, was the richest herring catching area. Nefyn beach, regarded
as very safe with gently sloping sand, had a narrow tidal range and boats
could be launched from the beach at almost any state of the tide. Double-
ended boats that were easy to launch from open beaches were preferred and
although sails were widely used, for herring fishing oars were preferred.

Nets were set in two ways, depending on where fishing took place. At Y
Gamlas in Nefyn Bay both ends of the net were anchored, while off Nefyn
Point at Y Swangins, only one end of the net was anchored, the other end
was buoyed and allowed to move with the tide. Not all those who fished
at Nefyn could afford a net and at the beginning of the twentieth century
a man at Nefyn acted as a net merchant. He loaned nets to fishermen and
as payment for his services he took a quarter of all the catches that the
particular fisherman obtained with the net for the duration of the loan.

Repairing nets on Nefyn beach (Photo – Gwynedd Archives)

When there was a new moon, it was believed that the herrings came close inshore and the fishermen spent some time in *Y Gamlas*. Without a moon, the herrings were much further out; hence *Y Swangins* was the main fishing area for a period. The type of herring net used at Nefyn was a fine meshed net, with a 1-inch mesh, usually purchased from netting firms in Bridport or Lowestoft. A standard net had to be adapted to local conditions; it had to be cut and selvedged and floats and weights were fixed. Herring nets were usually about 50 yards or 22 *gwrhyds* long and depth varied from about 6 feet to 10 feet. Stone weights were placed at the bottom of the net; the bottom rope in the traditional net being of grass. Egg-shaped stones, each about 4 inches long, were picked off Nefyn head and a channel for attaching the nets was cut round each stone. This process was known as *nitsio*. The stone weights or *poitshis* were attached to the net by means of thin threads, perhaps of wool that would break very easily in a stormy sea. At least in those conditions an expensive net could be saved. Weights were attached every *gwrhyd* (approximately 2 yards or as far as a man with outstretched arms could reach) along the lower tant while cork floats were attached to the upper *tant*, again with approximately 2 yards between each float. Each net was also equipped with one or two anchors made of 1¼-inch iron by local blacksmiths. The anchor had a shank of about 4½ feet with 2-foot prongs. To fish, the anchor was first thrown out; the boat was allowed to drift the length of the rope attached to the net, the net was then thrown out and a second net, and possibly a third, was perhaps attached to that to provide a long wall of netting. The other end of the net was either anchored or allowed to swing, a leather buoy (*bongi*), about 2 feet in diameter being joined to the net for floating and as a marker. Each end of the net was attached to a rope 4 *gwrhyds* long. A cork buoy in the form of a cross, known as *bwy cledda*, joined net and rope together. The net could be inspected to see whether anything had been caught after about six or seven hours in the water. If a catch was made, the net was pulled into the boat, the catch removed, and the nets again reset. The nets could be set in stepped ranks in the fishing ground. Nets could remain in place for five or six days, but it was essential that they were inspected daily before dawn.

The proceeds of the catch were shared equally between the members of the crew with one share being allocated to the boat. Quite often, the fish-merchants (*croeswrs*) did not pay until they themselves had disposed of the fish. If there was a glut of fish, more than the travelling merchants could cope with, Nefyn herrings could be sold to a Pwllheli merchant, who often paid a very low price.

Each fishing village in Gwynedd had its own fishing limits and it was an unwritten law that no fisherman from a particular village could trespass upon the fishing grounds of another. Herring fishing could begin in a minor way with the new moon in August (*dŵr newydd* – new water) although tradition dictated that September was the right time to begin. Others believed that no one should venture out until after Thanksgiving Day, about the third week in October. Herring fishing continued until January.

Although Nefyn always had some full-time fishermen who fished the mackerel, lobster and herring seasons, most of the fishermen were part-timers. Some were quarrymen, others were merchant seamen. Some even came from other fishing ports to take part in the Nefyn herring season, thus for example, Aberdaron fishermen lodged at Porthdinllaen for the herring season.

Further west, Aberdaron boats were concerned with herring fishing between October and February, but due to the strong current in the area, the fishermen had to be very careful indeed where they set their nets. Most of the 14-foot boats were manned by two men, but due to the small size of the boats, the men could only take two nets each. They were anchored at both ends or possibly anchored at one end only with a heavy 30lb stone acting as an anchor at the other. The current was far too strong to allow the nets to drift with the tide. There were specific places where the Aberdaron fishermen could set their nets and they attempted to fish in sheltered locations well away from the currents.

Herring fishermen at Nefyn, Gwynedd c. 1910 (Photo – Gwynedd Archives)

At Pwllheli and Criccieth Bay, large nobby-type boats had replaced the smaller variety by the turn of the twentieth century. Each boat carried up to 35 nets, each 30-40 yards long. They were fixed together to provide a long drift net. Five buoys were placed on each individual net together with cork floats. Nets were never anchored; boat and net were allowed to drift with the current.

Moelfre

The east coast of Anglesey was regarded as being an especially rich herring ground in the nineteenth and early twentieth centuries. Fishermen from Amlwch, Porthllechog, Bull Bay and particularly Moelfre caught substantial quantities of herring. Nevertheless, herring fishing had virtually ceased by 1933. The fishing season usually extended from October to February but it was customary for some of the full-time fishermen of Moelfre to move towards the Isle of Man to fish at the end of the local herring season.

The boats that operated in eastern Anglesey were 20 feet to 30 feet in length and carried lug sails, although for herring fishing the oars only were used. Each boat was manned by a crew of four, and each crewman operated two nets. The fish was usually sold to merchants (*tsiecmyn* sing. *tsiecmon*) from Bangor and Llannerch-y-medd who then sold the fish around the countryside. The fishermen themselves sold a certain amount directly to the public and young local boys were often given the task of walking the countryside selling from door to door. By 1914 however, most Moelfre fishermen sold herrings directly to Liverpool merchants and were responsible for transporting casks of herring to Benllech railway station.

The fishing day at Moelfre started around 5 a.m. and it is said that a local inn sold far more beer before 4.30 a.m. than it did for the rest of the day. Casks, each one to take about 600 herrings, were placed on the seashore and these usually numbered twelve to each boat. After about three hours fishing the boats returned to the shore and the catch placed in the casks.

The nets themselves were usually 80 yards long and those used early in the season had 12 meshes to the foot and were purchased from Musselborough. In January, as the herrings became smaller, a net with a fine mesh was required. This had 13 meshes to the foot and was purchased from a Bridport net-maker. Each net was 120 meshes deep. Around Moelfre itself, both ends of the net were anchored. Two yards of rope (*rhaff angor*) connected the net to the anchor, while the buoy (*bongi*) that ensured the net floated properly was a complete sheepskin, thoroughly tarred. The nets themselves were weighted with stones at 3-foot intervals and floated with corks at 8-inch intervals. The method of fishing was by under-running where one end of the net was taken on to the boat and all the nets were gradually passed in over one gunwale and out over the other. The fish were removed during the process, without bringing the nets on board.

In Benllech Bay where the currents were not so strong, only one anchor to two nets was required while in Lligwy Bay, three nets could be attached to one anchor. The nets were always stepped, with approximately 10 yards in between each rank of nets.

Aber-porth

Herring were caught by boats from most of the creeks of Ceredigion, but Aber-porth was perhaps the best known of all the herring ports. Most of the fishermen that were concerned with herring fishing, and the hey-day of the trade had passed by the end of the first decade of the twentieth century, were part-time fishermen and the crewing of the herring boats depended very heavily on merchant seamen home on leave for the autumn and winter. When Aber-porth and the adjacent villages of Tre-saith and Llangrannog possessed their own coastal trading vessels, those ketches and smacks were laid up in the winter at Cardigan and their crews would be concerned with herring fishing.

Herring boats, Traeth y Dyffryn, Aber-porth

Aber-porth, Dyfed in the early 1900s

The boats used at Aber-porth were heavy 25-foot or 30-foot craft that demanded a crew of between five and eight. Although the boats were equipped with sails for mackerel and lobster catching, for herring drifting, oars only were used. Usually each member of the crew had as many as four or five nets, so that a boat carrying twenty-five or more nets was not unusual, Each net was 50 feet long and a number could be fixed together to form a long wall of gill-netting. Unlike the Gwynedd herring fishermen, the Aber-porth crews allowed the herring boats to drift with the current in search of herrings. A net with a buoy (usually an old can), was thrown out over the gunwale while the other end of the net was attached to a length of rope, which in turn was secured to the boat. The boat and net drifted with the ebb as far as Mwnt and Cardigan Island, returning to Aber-porth and hopefully catching herrings on the flood tide. This method of fishing was known as *drifio*.

The other method of fishing was with fixed nets as at Moelfre and Nefyn. This was known as *tranio* or *setin*. A net, perhaps 20 feet in depth, was equipped with cork floats (*bwyau corc*). At the bottom of the net the rope was fitted out with a series of small round pebbles obtained from Cwmtudu beach. Each stone, known as a *collten* (pl. *collte*), was wrapped in a piece

An Aberporth herring boat and its owner in the 1890s

of rag and the rag was tied to the bottom rope of the net with string. Each net would be equipped with about 100 stone weights. A pair of anchors was required to keep the net in place. Alternatively, a net could be equipped with one anchor only, the other end being free so that the net moved with the current. The preparation of a net was known as *meinio*.

Tranio was carried out at a number of specific locations in the bay at Aber-porth. Each one had a name – Trân Cribach, Trân Croes y Traeth, Trân Fath Garreg, Trân Dwr Nel, Trân Ogo'Fraith and Trân y Llety.

The method of sharing the catch was somewhat different in Aber-porth for the boat, the net and the owner would receive a share each while the remainder was divided equally between all members of the crew.

Aberystwyth

Aberystwyth was, as noted, an important herring port. The herrings were caught in drift-nets and the boats went as far south as New Quay and as far north as Porthmadog. The drift-nets used were 30 yards long, 24 feet deep and with a mesh of 1¼ inches, and were corked and weighted in the Aber-porth manner. The larger boats carried fifteen to twenty nets each and the smaller, six to eight nets.

> During the season, fifteen to twenty boats or more would sail in company to
> the fishing grounds in time to shoot their nets at sunset. After a boat had lain
> to her nets for an hour or so, the first net was brought inboard and if the catch
> seemed satisfactory all the nets were hauled aboard and fish and nets were
> dumped in the bottom of the boat. The boats were so small and the working
> space so restricted that no attempt was made to take the fish out of the nets
> until the boats had returned to port. This meant that the nets could only be
> shot once on each trip.[47]

At high tide, the herring boats could enter Aberystwyth harbour, but at low
tide the catch was landed on the open beach. The herrings were bought by
merchants who in the late nineteenth century sent the herrings by rail to
Liverpool or sold them locally, but in all cases the merchants had to be on
the beach or in the harbour to meet the returning boats.

Milford Haven

Reference has already been made to the rich herring fishery in Milford Haven,
around the villages of Llangwm and Hook, but this has always been far less
important than sea fishing from the town of Milford Haven itself. Although
Milford had been a centre of mercantile activity and of inshore fishing in
particular since the Middle Ages, the potentialities of the natural harbour
were not fully realized until the last quarter of the nineteenth century. A
company to develop the docks was formed in 1874, but the growth of the
docks and the town was slow and painful; there were many bankruptcies
and mishandling of the technical problems of construction. The delay in
building the docks had brought the town to a low state; many of the small
industries had been closed to make way for the new docks; the population
had declined to less than a thousand; hardly a house remained with a roof;
the people were half starved and weeds and thistles grew in the street.

In 1884, however, the docks were opened, but the hopes of making
Milford Haven an important transatlantic port were never realized. As a
contemporary newspaper report states, 'Milford must sooner or later resign
itself to the inevitable; give up its dreams of becoming an ocean port and be
content with the more commonplace though no less useful role of a Welsh
Grimsby.' The proximity of Milford Haven to the rich fishing grounds of
the Western Approaches contributed in no small measure to its success as a
fishing port in the early years of this century. Moreover, the construction of

the docks coincided with the decline of Brixham as a fishing port and with the decline of inshore fishing in south-west England. As a result, many Brixham vessels were based at Milford, fishing in the Irish Sea.

Of course, herrings only formed a small proportion of the fish landed at Milford Haven. Between 1896 and 1904 three ice factories were built there and in Neyland, replacing the natural Norwegian ice obtained from a hulk in the docks. In 1902, a new wharf for landing the mackerel and herrings was constructed in front of the sea wall, outside the dock, so that the drifters could land their catch and go to sea again when the dock gates were closed. The quay was 250 feet long and a market 350 feet by 35 feet was built especially to deal with the fish. Herring smokeries were built and until 1914, Milford was a boom town. It soon overcame the decline of the First World War; by the 1920s, the herring catches were enormous, and five smoke houses were required for kippering. It ranked fourth, after Hull, Grimsby and Fleetwood, as the most important fishing port in England and Wales. In 1932, for example, it had 108 trawlers with 150 others from other ports based there for eight or nine months of the year. In recent years, there has been a spectacular decline in the fortunes of Milford as a fishing port. By 1974, the number of fishing vessels had dropped to 14 and today no Milford-owned vessels are based on the port. Nevertheless, a few vessels from other ports use Milford as a base, largely for mackerel fishing, the mackerel being frozen aboard ship and exported.

In the hey-day of herring fishing from Milford Haven, described in some detail in a Ministry of Agriculture pamphlet in the late 1930s, 48 some of the herrings were caught in trawlers as a by-catch of other fish, but most were caught by steam or motor drifters. Since the drifters usually returned to port daily with their catches, bunker space was much smaller than in trawlers. In a drifter the hull is planned so as to check any tendency to go down by the head, even under a burden of 40 or 50 tons of herring which could be taken in a single haul. The Milford drifter fleet, operating much further away from the coast than the small boats of Cardigan Bay, adopted standardized methods of herring catching found amongst the fishermen of all herring ports of Britain.

A British drifter customarily works from 50 to 100 nets made up into a 'fleet'. The number may be unrestricted or as prescribed by the Herring Industry Board in accordance with the conditions of the fishery at a particular port. Each net is 50 to 55 yards long and is composed of cotton or synthetic fibre.

The net is suspended by a 'strop' about 2 fathoms long from a 'buff, 'pellet' or buoy. As it would be impracticable to haul direct on this wall of delicate netting, the bottoms of the nets are attached by 'seizings' two fathoms in length, to a stout manilla warp which under-runs the whole fleet of nets and takes the strain when hauling. Each individual net is attached to the warp during the process of shooting and cast off again by hand specially detailed for the job when hauling. Hauling is done by means of the capstan. Since the buoy rope or 'strop' is about 2 fathoms, a minimum depth of 12 fathoms is required to ensure that the gear does not foul the bottom. Thus, when manoeuvring in the vicinity of a steam drifter fleet, it is possible to go far towards guarding against damage by keeping to waters where depths of not more than 12 fathoms are shown. If it should be necessary to cross a fleet of nets this should be done at right angles half-way between two buffs. If possible the propeller should be stopped before the net is crossed. Damage done under such conditions even by heavy draught ships is comparatively slight.

Under ideal conditions of weather a 'fleet' of nets, acting like a sea anchor, leads straight from the drifter's bows so that several 'fleets' of nets may be shot closely parallel with one another. Under other conditions the nets may be 'set about' by the tide so that one or more overlapping bights may be formed by a single 'fleet'. Buffs by their colour not only denote ownership of the nets but, by being painted one-quarter, one-half or three quarters white, indicate the position of each section of the 'fleet'.

Foreign drifters frequently work coarser nets than are used by British vessels and often employ a greater number of nets so that the 'fleet' may extend to three miles from the drifter. Dutch vessels, for example, when using their own type of gear work the net below the warp and their buoys consist of small wooden barrels. If the drift nets of British vessels and those of foreign vessels using the coarser net suspended below the warp become foul of one another the British gear, on account of its lighter construction, usually suffers.[49]

The Milford herring fishery is now virtually dead and the port has ceased to be of value in the fishing industry in general. There have been many changes since a correspondent to the *Pembrokeshire Herald* wrote in 1912, 'the fish trade is Milford's sole industry and everything and everybody in the town depend on it. Directly or indirectly between 1500 and 2000 people are engaged in it. The population of the town has doubled by means of it and thousands of pounds worth of house property has been erected as an outcome of its prosperity.'

In recent years herring fishing in the seas around Britain has been the subject of much complicated and restrictive regulation imposed by the European Economic Community. For some years with depleting stocks and over-fishing, herring fishing was totally banned, but in 1983 a limited amount of herring fishing was allowed, the fish being designated by the EEC as a 'Quota Species'. The seas around Britain were divided into a number of zones and only a certain amount of fish could be taken from those strictly delineated areas. Thus in Cardigan Bay and the Irish Sea (Zone VIIA) 4,660 tonnes of herring could be caught, while in an extensive region off the coast of south-west Wales and the south of Ireland (Zone VUC) the quota for 1986 was a mere 20 tonnes. The Bristol Channel together with the north and south Cornish and Devon coasts was limited to 250 tonnes. The complexity of the regulations was increased by allowing those vessels with so-called 'Pressure Stock Licence', to fish for herring:

> Quite often, these licences will have conditions superimposed on them, i.e. the main quota will be further subdivided and allocated to each vessel on a flat rate of say so many tonnes per day or week ... The opening and closing of the various herring fisheries in the Western Approaches and Irish Sea tends to be seasonal and also any management measures will be based on scientific advice given as a result of stock research... we can only hope that these management measures achieve the desired effect which is the conservation of the stock leading to greater amounts available to fishermen in later years.[50]

4 COCKLES AND MUSSELS

The collecting of cockles (*Cerastoderma* [*Cardium*] *edule*, Welsh *rhython*, *cocos* or *cocs*) and mussels (*Mytulis edulis*, Welsh *cregyn gleision*) is of great antiquity in Wales and there is extensive evidence from prehistoric and Roman sites that shellfish was important in the diet of early man. Throughout the centuries cockles and mussels, as well as other shore-living shellfish such as periwinkles (*Littorina littorea*, Welsh *gwichiaid*) and limpets (*Patella vulgata*, Welsh *llygaid meheryn* or *llygaid y graig*) provided the essential protein in the diet of coastal people throughout Wales. Shellfish occurred everywhere on the coast and were especially profuse along rocky stretches. They were very easy to collect and the gatherers of shellfish required no elaborate equipment, tools or great expertise to amass substantial quantities. To those prepared to venture to the sea-shore, no cost at all was involved in harvesting vast quantities of shellfish, that were very easily prepared for consumption in the farm or cottage kitchen.[1]

The industrialization of south Wales and the great increase in the population of the industrial valleys led to an ever-increasing demand for cheap food, with the result that the gathering of shellfish, cockles especially, developed into a commercial enterprise of considerable importance. Carmarthen Bay, in particular, was a major centre of shellfish gathering. In the estuary of the Tywi, said one nineteenth-century observer,[2]

> ... a large cockle fishery is carried on all the year round. This gives employment to 150 families, whose earnings are about fifteen shillings a week, in addition to the fish consumed by themselves. The value of the cockles and mussels taken near the mouth of the Tywi and up to Ferryside, is estimated at about £6,000 a year. Further east there are also cockle fisheries at Pen-clawdd and near Llanelli, in which some 500 families find employment and the cockles and mussels taken are valued at over £15,000 a year.

During the last half of the nineteenth century, cockle sellers from such places as Pen-clawdd, Llansaint and Ferryside became regular attenders at urban markets in south Wales, while dozens of cockle sellers with baskets on their heads and arms sold their products from door to door in all the industrial valleys.

In north Wales, the growth of Liverpool and of the Lancashire and Yorkshire conurbations provided a ready market for the shell fishermen of the Conwy in particular. The development of the railway in the mid-nineteenth century provided an easy means of transporting fresh sea-foods from the villages and towns of north Wales to the industrial cities. Before the arrival of the railway, most of the mussel gatherers on Conwy were concerned with finding pearls from the mussels. The ease of transporting perishable fish by railway contributed on the one hand to the decline of pearl fishing but on the other to the increase in mussel fishing for human consumption. Before that, the fish were put in a tub and stamped with the feet until they were reduced to a pulp. Water was added; the animal matter floated and was used as food for ducks, while the sand, particles of stone and the pearls settled to the bottom.

The cockle is a bi-valve mollusc, almost circular in outline. The beak or *umbo* of each valve is prominent and rounded from the apex of the shell to the crenated edge. In colour, cockles vary from a reddish-brown to yellow, the actual colour often depending on where they occur on a particular beach. At Llanrhidian sands in the Burry Inlet, for example, which is the main cockle-gathering area in Wales, the cockles that occur near the low tide mark are yellow in colour and are usually sold 'in the shell'. Those further up the beach near the high tide mark are grey and orange with black patches and since they look less appetizing, they are sold after boiling and shelling and are described as 'boiling cockles' as opposed to 'shell cockles'.

Only the expert cocklers [said one observer], know when a bed is ripe to be worked. For a long time beforehand, perhaps for years they observe and test the 'seed' beds of young cockles: these are very densely populated and may contain 3,000 cockles to the square yard. After its brief swimming stage, the cockle settles in the sand. When about a twentieth of an inch long it lives on minute plants and organisms in the seawater and can move through the sand by means of its strong muscular foot. It usually adds two rings to its shell in the first year and one a year after that. At three years of age it is about an inch across and... it may live six years.[3]

The breeding season for cockles is in April, May and June, but cockle gathering is carried out throughout the year without a break.

The edible mussel is, like the cockle, a bi-valve mollusc and has a bluish-black, almost triangular shell. It is very common and can be found almost everywhere along the Welsh coast. Nevertheless, the greatest concentration of mussels is in river estuaries, where they occur both on the shore and below low-water mark. The principal mussel fisheries in Wales today are in the Conwy estuary, but substantial quantities are caught elsewhere, more especially in Carmarthen Bay near the villages of St Ishmaels and Ferryside and, until recently, in Porthmadog harbour in Gwynedd. Mussel farming is carried out on a considerable scale in the Menai Straits, where seed mussels are laid for future harvesting. The best mussels occur in those places where salt water is replaced by fresh water at regular intervals. Those mussels that live below low-water mark and are therefore always covered with water, grow far more quickly that those occurring on rocks and banks that are exposed at each low tide. Mussel fishing is an autumn and winter occupation for in the spring and summer months mussels do not possess much flesh, for the condition of the flesh varies according to the state of the gonad, the reproductive organ. 'It improves throughout the autumn and winter as the gonad fills with reproductive cells and occupies more and more of the flesh. It is at its best just before spawning, after which it becomes poor.'[4] Since spawning takes place in the winter and early spring, few mussels are gathered between April and October.

Cockle gatherers on Ferryside beach in Carmarthen Bay c. *1910: preparing for a gathering Session*

Loaded donkeys leaving the beach for the village of Llansaint where most of the cockle women lived

Cockle gathering

Cockle gathering, despite a recent decline in the number of gatherers, still remains an important occupation in Carmarthen Bay, especially so in the Burry Inlet on the east side of the Bay. It is an occupation that is still carried out by women to a considerable extent and most of the gathering is done at low tide on Llanrhidian sands by the inhabitants of three north Gower villages: Pen-clawdd, Croffty and Llanmorlais. A certain amount of gathering was done until recently on the north shore of the Inlet around Llanelli, while further west the estuaries of the Tywi, Taf and Gwendraeth were fished by the inhabitants of such villages as Llansaint, St Ishmaels, Ferryside, Laugharne and Llansteffan. In this western sector of the Bay, the industry is now of minor importance and most of the gathering is done on Llanrhidian sands. In 1966 one observer noted that 'annual landings from this relatively small area are exceeded only by those from the vast areas covered by the cockle fisheries of the Wash and the Thames estuary. In 1953 and 1954, the quantity of cockles landed in all three areas was very similar: those from Burry Inlet constituting one third of the country's recorded total'.[5]

65

There has certainly been a great decline in cockle fishing in recent years. In 1910, for example, Bulstrode estimated that at Pen-clawdd there were 250 cockle gatherers with 150 operating in the Ferryside, Llansaint district, 50 at Laugharne, 12 at Llansteffan and 50 at St Ishmaels.[6] In the Burry Inlet women from other villages outside Pen-clawdd, Croffty and Llanmorlais were engaged in cockle gathering. Many came from Gowerton and Loughor and others from hamlets such as Wernffrwd and Penuel. In 1921, Wright maintained that the number of people engaged in cockle gathering numbered between 100 and 200 in winter and 200 to 250 in summer.[7] 'They were exclusively women from Pen-clawdd and neighbouring villages, except during periods of local unemployment such as due to stoppages in the coal industry, when a number of men took up cockle gathering.'[8] Until recently, the number of cockle gatherers operating in Carmarthen Bay has varied between 80 and 100. In 1965, for example, 80 gathered at Llanrhidian sands; 12 from Llansaint worked the Ferryside and Gwendraeth beds and two from Laugharne fished Llansteffan beach. In 1971, 76 licences were issued by the South Wales Sea Fishing Committee for the Llanrhidian district, 12 at Ferryside and 10 on the Llanelli shore of the Burry Inlet. In 1976 for the whole of Carmarthen Bay, only 45 licences were issued and many of those licensees never ventured on the sands to gather cockles. Llansaint, which was an important cockling village in the past, in 1977 had only two cockle gatherers working. In 1985 the total number of licences issued was 52, while in 1991 a total of 46 licences were issued. Many families who were dependent on the sale of cockles in the markets and towns of south Wales have ceased to take an active part in the industry.

A number of reasons have been cited for this gradual decline. Pollution from industrial undertakings and considerable extension in the growth of tough clumps of spartina or cord grass resulting in a poor spatfall are partly to blame, while hydrographic alterations in the course of the River Loughor have certainly contributed to a decline in the cockle population. Forty years ago changes in the courses of the Rivers Taf and Tywi resulted in the washing away of cockle beds in the north-western part of Carmarthen Bay. The cockle gatherers themselves believe that the great increase in the population of mollusc-eating oyster-catchers has been the prime factor in the decline of the industry. Oyster-catchers do eat vast quantities of cockles, indeed it is estimated that between December 1961 and March 1962, for example, six hundred million were devoured by the birds during observations in the Burry Inlet.[9] In recent years there has been a campaign for the planned culling of oyster-catchers on the cockle beds, despite the objections of conservation groups.[10] There is also a strongly held belief among the cockle gatherers of Pen-clawdd that cockles migrate. In the early 1940s, for example, few cockles were being gathered at Llanrhidian sands and it was believed that all the cockles had migrated to Ferryside and the mouth of the Tywi. The cockle women of Pen-clawdd had to travel daily, often as early as 4 a.m., through Gowerton, Llanelli, Cydweli and St Ishmaels to reach the cockles.

Cockle gathering on Llanrhidian sands, Burry Inlet, 1980: exposing the cockle beds

Cockle gathering at Llanrhidian: 'riddling' the cockles

There was a strongly held belief too, that if two people quarrelled on the beach while gathering cockles, then the fish would migrate immediately and cause great inconvenience to the community.

Despite the limitations on the number of cockles that may be gathered in anyone day, there has certainly been over-fishing in the Burry Inlet especially. In the early 1920s donkeys were used exclusively by the cockle gatherers for transporting fish in bags from the beach and each cockle gatherer was limited to 2 or 3 cwt. by the weight that each donkey could carry. The 200 or more donkeys at Pen-clawdd were a source of considerable disturbance in the villages. In the 1920s horse-drawn vehicles were not known on the beaches. Generally, says Matheson[11] 'the gatherers make use of donkeys to carry their cockles from the beds to the foreshore, probably because the hoof marks of so light an animal do far less damage to the beds that would a cart drawn by a horse and fully loaded'. Gradually the donkey disappeared from Llanrhidian sands to be replaced by horse-drawn, rubber-tyred flat carts that could carry far more than a couple of hundredweight of cockles. So many cockles were being taken from the Burry Inlet that the Fisheries Committee were forced to set a limit in 1952 on the number taken in anyone day. 'With the use of lorries on the Llanrhidian sands, cockles were being gathered indiscriminately during the summer months, to the detriment both of the winter fishery and of that of the following early summer.'[12] In 1967 each gatherer was allowed to collect 10 cwt. in a day; in 1968 it was reduced to 7 cwt. and in 1972 it was reduced to 5 cwt. with a further reduction to no more that 2 cwt. in 1974. There are also legal limits as to the size of cockle that may be picked:

> In 1921 there was a by-law regulating the size of cockle, that could, legally, be removed from Llanrhidian sands. The legal minimum size was ¾ inch, i.e. no cockle should have been taken which would pass through an aperture ¾ inch square, but since 1914 this by-law had covered only 'shell cockles, i.e. despatched in the shell and had been relaxed to allow the taking of boiling cockles down to ⁵/₈ inch. This situation continued until 1953 when the South Wales Sea Fisheries Committee tired of the confusion and administrative difficulties resulting from the M.A.F., and as a result standardised the minimum size at 11/16th inch. Later in 1959, as a result of fresh scientific evidence, the minimum size was raised to ¾ inch.[13]

In the 1960s the quantity of catch taken from the beaches, especially on the Llanrhidian sands, increased spectacularly with the sharp rise in the price of cockles. The decade also saw the almost total replacement of the donkey as a means of transport by a fleet of horse-drawn carts. With the closure of coal-mines and a number of tin-plate works in the region, some men became regular cockle pickers and Llanrhidian sands were no longer the exclusive preserve of the women of Pen-clawdd. In 1969 as much as 90,264 cwt. of cockles were gathered in Carmarthen Bay. A total of 79,872 cwt. were collected in 1970; 67,424 cwt. in 1971; 84,100 cwt. in 1972; 59,865 cwt. in 1973. In 1974 with the depletion of the cockle population only 12,693 cwt. were gathered and in 1975 the quantity had declined even further to 4,864 cwt. There was overwhelming evidence, despite legal limits of size and quantity directives by the Fishing Authorities, that there was substantial over-fishing in the Burry Inlet. But in the 1970s the cockle beds did recover considerably although the number of gatherers had declined. In 1985, 52 people were licensed to gather cockles in the Burry Inlet but by 1988 the number had increased to 66.

In the 1980s, all was well in the cockle gathering industry of Carmarthen Bay and 69 licensed gatherers were working in Burry Inlet. The value of the harvest approached £250, 000 with 65,000 hundredweight landed.

Since then a severe depression has entered the industry. The threat of oil pollution and a mysterious disease amongst the cockles, that has restricted their growth, has meant that most of the small fish have been returned to the sands. Only 41 gatherers are in part-time work in 2009. The industry is strained further as gatherers, mainly from North Wales, together with European and Chinese unlicensed operatives, invade the beaches, often creating many a heated argument upon the sands. All is not well in the Carmarthen Bay.

Cockle landings in the Burry Inlet

Year	Cwt.	Value
1972	84,100	£105,292
1973	59,770	£84,711
1974	12,693	£17,597
1975	4,864	£7,291
1976	8,060	£12,090
1977	12,448	£23,524
1978	13,833	£27,666
1979	17,938	£80,348
1980	25,022	£125,110
1981	39,600	£201,000
1982	40,428	£202,420
1983	31,644	£158,220
1984	36,229	£181,145
1985	39,063	£195,315
1986	36,008	£180,040
1987	65,163	£365,484

(*Source:* Figures issued by the South Wales Sea Fisheries Committee.)

A Pen-clawdd cockle cart on the way to the cockle beds on Llanrhidian Beach

The cockle gatherers of the Burry Inlet start from their homes on the ebbing tide and this may well be before dawn breaks. Fleets of rubber-tyred carts make their way through the village streets and across the inhospitable, windswept salt marshes towards the beach, a distance of two miles or more from the villages. Occasionally the convoy has to wait at Salthouse Point before the sea has receded enough to venture on the sands. The journey across the beach may be one of considerable peril, especially at night or in foggy weather. The rapidly flowing streams or 'pills' that run across the beach may be treacherous and can spell doom to horse, cart and cockle gatherer.

Each woman after arriving on the beach selects her sections and begins to gather cockles. There are no set rules as to where the cockle women pick the shellfish: anywhere will do as long as it is not in front of or behind another picker. Cockle picking is entirely a hand process and the picker is armed with a small knife with a curved blade about six inches long that is used in the right hand to break the surface of the sand, thus exposing the cockle beds. The knife is known as a *scrap* in Pen-clawdd and as *y gocses* in Llansaint. The handle of the tool is either bound with twine or a piece of rag to make it more comfortable for use. In the cockle gatherer's left hand is a hand rake (*cram* in Pen-clawdd, *rhaca gocs* in Llansaint). This is used for drawing together the cockles into heaps and for this the rake is usually transferred to the right hand. The heaps are then put into a sieve or riddle that is placed slightly behind but between the legs of the gatherer. It will contain sand, empty shells, as well as live cockles. The sieve with a mesh of ¾ inch is then shaken backwards and forwards and from side to side, so that all undersized cockles fall through the mesh. The sieve has a diameter of about 18 inches and is wired like an oblong-meshed builder's sieve; that is generally preferred to square-meshed sieves. Most of the equipment required by the cockle gatherers is locally made. The *scrap* is usually shaped from an old sickle and the blade turned to an S-shape under the heat of a fire, while the *crams* are made by blacksmiths; each one made specifically for an individual cockle woman. Although ordinary builder's sieves are used today, until about thirty years ago the sieves were made by the cockle women themselves or by specialized sieve-makers (*gwilodwr/gwaelodwr – bottomer*) in Llansaint. Baskets for carrying cockles and for washing them were, until recently, produced by village craftsmen,[14] while the sacks used for carrying the cockles from the beaches were obtained from local farmers.

After sieving cockles, the shellfish are washed in a convenient pool or 'pill' and the harvest placed in sacks ready for transporting back to the village. When donkeys were used for transport, the sacks had to be sewn with needle and twine to prevent the fish from spilling out on the journey back from the beach. Two bags of cockles were slackly sewn together and slung over the donkey's back and the gathering equipment was then placed on top of the sacks for transporting. Although donkeys were still used by the Llansaint women for as long as cockle gathering was practised, mainly because the beaches of the area have much soft sand (*trath shinc*), at Pen-clawdd carts are used exclusively for transport. A few two-wheeled carts were used by the Llansaint women for transporting cockles from Ferryside beach in the 1920s, but they were soon abandoned as unsuitable.

The life of a Carmarthen Bay cockle woman is not an easy one, for she has to be out on the beach in all weathers and at all times of the year. Constantly bending over a cockle bed is tiring work and the shaking of the sieve to get rid of sand, empty shells and undersized fish demands considerable energy. The old cockle women always wore many layers of clothes: an elastic-edged cap or 'pixie' covered with a shawl (*shol drath* – beach shawl) tied around the neck ensured that the head was protected. A sack apron was considered essential and this was inserted between the legs and pinned at the back and had the appearance of a pair of trousers. Woolen cardigan, one or two flannel shirts, woolen underpants, a shirt and one or two flannel petticoats with knee boots completed the ensemble.[15]

A Pen-clawdd cockle seller in Swansea market in the 1930s. (Photo – M.L. Wight)

In the 1950s an attempt was made, especially in the western part of Carmarthen Bay, to introduce a hand dredge that was expected to speed up the gathering process considerably:

The dredge is dragged towards the operator who moves backwards, pulling with sharp movements alternatively one side of the handle, then the other, whilst keeping the toothed blade below the surface of the sand. When used in a flowing stream, where cockles often accumulate, the sand is washed away through the diamond mesh, leaving clean cockles in the dredge. On the surfaces with only a limited film of water, the dredge is up-ended and rocked vigorously from side to side, when cockles are again freed from the sand. The designer and one or two gatherers claimed great success for this method during the late 1950s, particularly on beds containing larger cockles.[16]

Dredges have not been in general use since about 1958 although an attempt was made to reintroduce the dredge to the Llanelli cockle beds, on the north shore of the Burry Inlet in the mid-1960s.

The best cockles come from sandy, rather than muddy stretches of beach and most of the gathering is done on Mondays and Tuesdays with a certain amount of gathering on Wednesdays. On Thursdays cockles are boiled and prepared for the market on Fridays and Saturdays. Thursday was particularly important traditionally as a cockle gathering day on Llanrhidian sands[17] and 'on Fridays the amount gathered is reduced to about a quarter of the previous daily quantities, because many of the gatherers are preparing for or taking part in weekend markets'. There is no gathering on Saturdays and Sundays. During a week in June 1973, to quote one example, it was observed that on a Monday there were 62 cockle gatherers operating on the morning ebb; on Tuesday 65, Wednesday 38, Thursday 10, Friday 4. Usually an officer of the South Wales Sea Fisheries Committee is present on the beach to ensure that all the regulations regarding the size of cockles and the quota gathered by each cockle woman are strictly carried out.

Today, about 25 per cent of the cockles gathered on Llanrhidian sands are sold in the shell, but at Llansaint most were sold without cooking. All that is required to prepare the cockles for market is to wash them in clean water. In hot weather the cockle women believe that it is unwise to handle cockles excessively and washing is perfunctory to say the least. At Penclawwd most of the cockles are cooked and shelled ready for the market, usually in communal 'boiling factories'. Traditionally at Llansaint a coal fire was lit on open ground or at the bottom of a garden and the cockles were boiled in a metal pan, but it was important that the fire did not touch the bottom of the vessel. Very little water was added to the cockles, since they contained enough water themselves. The shellfish were then sieved while still hot, so that the flesh was separated from the shells. They were then scalded and washed in clean water. They were again sieved, washed and boiled for a very short time in slightly salted water and then spread evenly

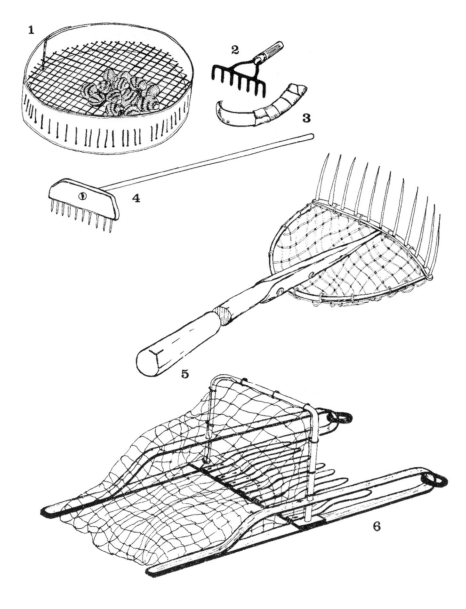

1. *Cockle sieve – Croffty, West Glamorgan*
2. *Cockle rake(cram) – Pen-clawdd, West Glamorgan*
3. *Cockle knife (scrap) – Pen-clawdd, West Glamorgan*
4. *Long-handled cockle rake – Croffty, West Glamorgan*
5. *Mussel rake – Conwy, Gwynedd*
6. *Cockle dredge – Laugharne, Dyfed*

Inside a Pen-clawdd cockle factory where the fish is shelled, sorted and boiled

on a wooden plank to cool. At Llansaint shell cockles were delivered to customers from sacks carried on the donkey's back, but the cooked cockles were carried in baskets. Most were sold from door to door, usually to regular customers.

Bulstrode gives a detailed picture of the Burry Inlet in 1911:

Along the foreshore, there are to be seen numerous extemporised fireplaces, banked up with sand on three sides to protect them from wind and open in front. Upon these fires the cockles are placed, contained in a large iron or tin cauldron, a small quantity of water having been first poured into the saucepan to prevent scorching. As the cockles gape owing to the passage of steam from below, the sea-water contained between their shells is liberated and thus the cockles may be said to be literally boiled in their own liquor, this liquor being at the end of each boiling decanted almost entirely into another receptacle near, to be used subsequently for the process of washing the cockles ... The cockles are kept on the fire until the water is actually boiling, when they are emptied into another receptacle and a fresh batch of living cockles placed in the cauldron. The boiled cockles are then riddled over a table, the soft parts passing through a sieve, the shells remaining behind. From time to time the cockles are washed in their own liquor and at the end of the process they are taken, in a basket either to a neighbouring stream, to a spring, on the foreshore or to a pump or tap... At the end of a day's work at Pen-clawdd, there may be seen numerous cockle-wives sitting by the side of a rivulet which traverses the sands, awaiting their turn to wash their boiled cockles. The cockles when ready for market are wrapped in white cloths, placed in baskets or small wooden tubs and taken to the neighbouring market towns by the cockle wives, who may be seen in the early morning on the railway platforms with basket or tub deftly poised upon their heads. On Saturdays, a special train leaves Pen-clawdd for Swansea about 7 a.m., which is known as 'the cockle train'.[18]

In the western part of Carmarthen Bay, when the cockle gathering industry flourished in such places as Laugharne and Llansteffan, boiled cockles were usually heavily salted with 28lb. of salt being added to every 60 quarts of cockles.

Today in the Burry Inlet there is no cockle boiling in the open air, but there are a number of communal plants, 'at each of which some 20 to 30 gatherers may pay to cook their cockles'.[19] As a rule they are owned by cockle gatherers who themselves are usually retailers and distributors of cockles. There are also a number of privately owned plants, where only the cockles of one individual cockle gatherer may be prepared.[20]

The method of preparation adopted in Mr Selwyn Jones's factory, a Nissen hut on the saltmarsh, is typical of the technique adopted generally in the Pen-clawdd district. The cockles are trasferred from sacks into perforated buckets or circular baskets and placed two at a time in a small steaming boiler. With steam entering the boiler the liquid in the cockles is forced out through a drainage hole at the bottom of the boiler. After cooking for six or seven minutes the cockles are then transferred to a revolving cylindrical wire screen powered by electricity. The flesh falls through the wire mesh on to trays below and the shells drop down to the lower end of the cylinder, which is at an angle of 45 degrees, through a chute that leads to the outside of the building. The shells are usually sold as poultry grit. The meats collected in the trays are then placed in baskets and washed in a series of containers filled with fresh water. The cockles are swirled around with the bare forearm so that sand and other impurities are removed. The cockle meats are then packed in baskets ready for sale.

The main outlet for Pen-clawdd cockles is Swansea market, although some of the women regularly attend other markets such as Llanelli, and Carmarthen. J.M. Thomas in 1949 described the marketing as follows:

> Dozens of women still sell door-to-door in all parts of South Wales, since it is a point of honour to satisfy old customers. Early on a Saturday morning there is an exodus to towns and villages in the Swansea, Neath, Rhondda and Gwendraeth valleys; to Carmarthen, Cardiff, Newport; one girl goes all the way to Bath. Transport however, has helped to commercialise the distribution and most of the cockles are sent away in bulk to London and all the main towns and cities of Britain. Bottling cockles has been successful ...[21]

Of course, attendance at market and vending from town to town still remains important despite modern methods of distribution:

> A generation ago, the Saturday morning train from Pen-clawdd to Swansea used to be a colourful sight: almost the whole female population of the villages would crowd on the platform dressed in Welsh costume – red and black striped dresses, black and white check aprons, grey plaid shawls and close-fitting cockle bonnets, with a starched white frill peeping out from under the brim – and the white cloths over their tubs and baskets would shine like so many fresh mushrooms. After selling their cockles they would return on the 2.30 train from Swansea.[22]

Before the railway came to Pen-clawdd in 1863 the women used to walk the nine miles to Swansea market in bare feet with tubs of cockles balanced on their heads and it is said that a stream on the outskirts of Swansea known as *Yr Olchfa* (the washing place) was given its name by the cockle women because it was there that they washed their feet before putting on their boots and entering the town.

Mussel gathering

For centuries the most important centre of mussel gathering in Wales has been the Conwy estuary, and in 1974 it was estimated that the total value of mussels landed at Conwy amounted to £40,000 for the year. In 1984 about 28 full-time fishermen were engaged in mussel gathering; a considerable decline from 1939 when the number of fishermen amounted to 75, including 8 full-time women gatherers. The depletion of mussel stocks in the Conwy estuary has now reached serious proportions, with the result that there has been a programme of artificial planting of Anglesey mussels in the estuary in an attempt to stem the decline of the industry. There has also been a considerable decline in the mussel-gathering industry in Porthmadog harbour, where in 1974 the estimated value of shellfish landed amounted to £20,000. Near Bangor, however, mussel farming, where mussel spat is laid on the sea bed to be collected later when mussels are of marketable size, is of increasing importance. Mussel gathering that was widely practised in such places as Pwllheli, Barmouth, Aberdyfi and the Menai Straits before 1939 has virtually disappeared. A certain amount of mussel gathering is still practised in Carmarthen Bay, usually in association with cockle gathering.

In the nineteenth century, mussels in the Conwy were usually gathered for their pearls, and the history of pearl fishing in the Conwy dates back for many centuries. Spencer in this *Faerie Queene*, for example, wrote:

> Conway, which out of his streame doth send
> Plenty of pearles to decke his dames withall.[23]

A local historian in 1835 gives details of the pearl fishing at that time:

> There are two kinds of mussels found in the Conwy from which pearls are obtained: *mya margaritifera* (*cragen y diluw*) and the *mytulis edulis* (*cragen las*). Those of the former species are procured high up the river, about Trevriw and pearls scarsely inferior to the oriental ones are occasionally found in them ... The other variety, the *cragen las*, is found in abundance on the bar at the mouth of the river and great quantities of the mussels are daily gathered by numbers of industrious persons. At ebb tide, the fishers, men, women and children may be observed busily collecting the mussels, until they are driven away by the flood. They then carry the contents of their sacks and baskets to *Cevnvro*, the northern extremity of the marsh, where the mussels are boiled: for this operation there are large crochanau or iron pots, placed in slight huts;

or rather pits as they are almost buried in a vast heap of shells. The fish are picked out and put into a tub and stamped with the feet until they are reduced to a pulp; when, water being poured in, the animal matter floats, which is called *solach* and is used as food for ducks, while the sand, particles of stone and the pearls settle in the bottom. After numerous washings, the sediment is carefully collected and dried and the pearls, even the most minute, are separated with a feather on a large wooden platter.[24]

By the early 1880s pearl fishing on the Conwy had ceased[25] but mussels were being gathered in considerable quantities for human consumption. In its heyday the pearl fishery was of some importance in Conwy and Samuel Lewis in his *Topographical Dictionary of Wales* (1833), stated that 40 persons were employed in the fishery producing about 160 ounces of pearls a week. The boiling of mussels was carried out in especially constructed 'pearl kitchens', each kitchen being '5 feet square and 6 to 7 feet high – constructed of wattle and gorse, a hole being left for smoke to escape'.[26]

Mussel dredging in the Conwy estuary: the boat

Mussel dredging: scraping

Mussel dredging: collecting

Pearls were collected in a number of the Welsh rivers in the nineteenth century, among them being the Cleddau, Ely, Teifi (at Maes-y-crugiau), Tywi (at Llanarthne) and Taf.[27] Itinerant Scottish pearl fishers paid annual visits to this last river during the decade 1926-36 and fished for pearls in the Whitland district in July and August when the water was low.

As far as commercial gathering of mussels for human consumption is concerned, the fishing season in the Conwy estuary extends from September to April. Gathering may either be done on the shore, usually with the aid of a small knife (*twca*), a process that is described as *hel ar y lan* (collecting on shore), or the fishing is done in deep water from a boat, using a long-handled rake or *cram*. This method of collecting is described as *codi o'r dwfn* (lifting tram the deep). Today eight boats are engaged in the fishing and the fishermen, licensed by the local authority for no annual charge, are members of four families. Three of those families – Jones, Hughes and Roberts – have been Conwy fishermen for many generations but the fourth, the Cravens, have only been fishing there since about 1920. Each family group has its traditional point of embarkation – the Jones's at the Custom

House, the Hughes's at Cei Main and the Roberts's at the Janus Arms. Each boat as it leaves Conwy may have two fishermen who will gather 'from the deep' and perhaps two or three women who will disembark in the estuary to gather mussels on the shore. The fishing is done for a period of about four hours at low tide and departure from Conwy is planned in such a way that the boats will sail downstream on the ebb tide, returning to the quayside on the flood. Regulations state that there must be at least four feet of water at low tide over the mussel beds, while the rake used for gathering must not exceed a width of three feet. Each mussel bed in the estuary is named – Y Popty, Cae Gonwy, Sgiaps (scabs), Y Marfa, Y Men, Westras (oyster) and Cerrig Duon.

Having reached the mussel beds, the boat, usually a 14-or 16-foot rowing boat with outboard engine, is anchored fore and aft over the mussel bed. More often than not a single anchor is thrown out over the bow and a bucket or basket over the stern, so as to keep the boat in place over the mussel bed and in line with the flow of the tide. The fisherman standing in the boat takes the long-handled rake, with a pitch-pine handle as long as 30 feet, drops the prong end into the water as far from the boat as the fisherman can reach. The rake has ten or a dozen long prongs, 8 or 10 inches long and on the back of these prongs is fitted a net or purse. With the prongs resting on the bed of the river the pole is placed on the shoulder whilst the fisherman draws the rake towards him over the bed, all the time exerting a downward pressure on the handle to remove the mussels attached to the rocks. This is back-breaking work and when the handle of the rake is nearly vertical, it is turned over so that the mussels caught between the prongs are transferred into the attached net. The rake is then drawn up, hand over hand, shaken and the catch dropped into the centre of the boat. The process is repeated until the boat is heavily weighted with mussels ready for the journey back to Conwy Quay on the flood tide. Each fishing boat usually carries two mussel rakes; the one with a handle of about 12 feet long and another measuring anything from 18 feet to 30 feet in length. In some cases for the longer rakes, it is necessary to bind two lengths of pitch-pine together to provide the required length of handle, but whatever size of rake is required the handle must be very smooth. Usually the diameter ranges from 2 inches at the bottom to 1½ inches at the top and a rake produced at a Conwy boat-yard and blacksmith's shop should last for four or five seasons. When not in use the rakes are left on the beach at Conwy, usually in the vicinity of the slipway or town wall.

Miss Gwladys Roberts of Conwy gathering mussels from the rocks on the sea-shore

The only instrument used by the shore pickers, who are usually dropped at their picking station in the estuary by the family boat, is a small knife known as a *twca*. This has a spoon-shaped blade made by a local blacksmith to the individual specifications of a mussel gatherer. The length of blade according to one gatherer should be equal to the 'length of the inside of the middle finger'. The mussels are prised away from the rocks with the *twca* held in the right hand and three or four mussels at a time are pulled away from the rocks with the left hand, before they are transferred to a willow or wire basket placed conveniently near the gatherer. Usually three basketfuls constitute a bagful. One advantage that shore gathering has over fishing from the deep, is that the gatherer can select the most suitable mussels, rejecting those that are too small or too large. Shore gathering is certainly arduous work and the gatherers, many of them being women, usually bind their hands with rags to prevent them from being cut with sharp mussel shells. They are usually dressed appropriately for a cold windswept, estuarine shore, In the past they wore ankle-high clogs, a woollen shawl (made by the Trefriw woollen mills) over the head, a woollen skirt, coat and two or three flannel petticoats. Men wore layers of underclothes and a black tweed cap that came over the ears and fastened under the chin.

It was customary for the shore gatherers, once they had collected a bagful of mussels, to walk the three or four miles back to Conwy and sort out the mussels on the quayside. They would then go to their respective homes for a meal, walking back to the quayside in time to meet the men, who returned with their boats on the flooding tide. The next four hours were then spent in sorting out the mussels gathered from the deep. Sorting again, is always done in family groups and the women of each family are expected to help with sorting although they have no part in gathering the mussels from the beds. The sorting is usually done in the boat and the sorted mussels are placed in baskets, washed in the river before being transferred into sacks that have to be carried to the purification plant in the shadow of Conwy castle.

The mussels are spread two deep on wooden grids on the bottom of shallow tanks. The collection of each fisherman is marked out by placing stones around a section of flooring. The mussels are then hosed thoroughly so as to cleanse the outside of the shells, sterilized sea-water containing chloride of lime is then pumped into the tanks and for one day the mussels open their shells and discharge the stomach contents, taking in sterilized sea-water instead. The water is then run off and a fresh supply admitted which is again run off after a day. The pattern of operation is that mussels placed in the tanks on Tuesdays are removed on Saturdays. All the mussels are finally sterilized in a weak solution of chlorinated water: they are bagged, sealed and sent away to market. The sterilized bags, usually taken by merchants to the Sheffield and Birmingham markets in particular, are fastened by one continuous string, so arranged that when sealed at the Ministry of Agriculture's purification plant, they cannot be opened without breaking the seal or cutting the string.

The mussel purification plant at Conwy

5 LOBSTERS AND CRABS

Lobsters (*Homarus gammarus*, Welsh *Cimwch* pl. *cimychiaid*) and the crabs (*Cancer pagurus*, Welsh *Crane* pl. *crancod*) are the most valuable fish caught along parts of the Welsh coast today. Nevertheless the commercial exploitation of the plentiful supply of lobsters and crabs is of no great antiquity in Wales although in the eighteenth and nineteenth centuries appreciable quantities for local consumption were caught off the rocky coasts of Pembrokeshire and the Llŷn peninsula. The lobster and crab were never popular items in the diet of Welsh people and it was the development of improved methods of transportation whereby the harvest of shellfish could be more easily taken to urban markets that really led to the growth of lobster and crab fishing on a commercial basis. In the five years 1902-1906, for example, a total of 13,826 lobsters were landed in Welsh ports, but that figure was exceeded in 1912 alone when no fewer than 14,637 lobsters were caught.[1] Crab landings, too, had increased from 15,505 between 1902 and 1906 to 30,021 in 1912.

Although the commercial development of lobster fishing was insignificant along most of the Welsh coast until at least the last quarter of the nineteenth century, the activity was of considerable importance on Bardsey at a much earlier date. 'Collecting of lobsters and crabs occupied most of the time of the inhabitants of Bardsey Island' said Bingley in 1800.[2] The pots were 'made of willow exactly in the shape of a wooden mousetrap with the cone inverted and the catch was sent to Liverpool by boat'. These lobster pots, like many of those used around the coast of Llŷn today, were fished singly and were weighted down with three stones attached to them on the outside. The regular boat service to Liverpool mentioned by Bingley continued throughout the nineteenth century until 1914 and was regarded as the lifeline for the population of Bardsey. There are few other references to the existence of a lobster fishery in Wales during the major part of the nineteenth century although Pennant in 1810[3] refers briefly to a certain amount of lobster fishing off the Llŷn coast by Pwllheli fishermen where 'traps for lobsters are made with pack thread like thref nets and baited with pieces of lesser spotted shark'. Holdsworth in his comprehensive survey of sea fishing in 1874,[4] and Buckland and Walpole in their 1877 *Report on the Crab and*

Lobster Fisheries of England and Wales, of Scotland and of Ireland[5] make
no reference at all to lobster and crab fishing in Wales, suggesting that this
category of fishing was not commercially developed at that time.

In 1884 a competitor at the Liverpool National Eisteddfod[6] again
mentions Bardsey and Llŷn as important centres of lobster and crab fishing
and 'everyone on Bardsey from the King of Bardsey downwards obtain a
livelihood from tilling the land and lobster fishing'. The winning essayist in
that same Liverpool Eisteddfod[7] notes that the north Pembrokeshire coast
too was an important centre for lobster fishing and both lobsters and crabs
occurred as profusely along the rocky coast of north Pembrokeshire as they
did around the cliffs of Caldy Island. At Dinas, near Fishguard, 'the Women
of Cwm yr Eglwys who own about half a dozen boats carryon a small
but profitable fishery in crabs, lobsters, etc. – most of the fish are disposed
of locally'. At a much earlier date, George Owen of Henllys was most
impressed with Pembrokeshire lobsters.[8] 'The lobster', he said in 1603, 'sett
whole on the table has its special qualities, he yeldeth exercise, sustenance
and contemplation; exercise in cracking his legges and clawes; sustenance
by eating the meate; contemplation in beholdinge the Curious work of his
compleate armour both in hue and workmanship.' Lobsters occurred widely
along the cliff-lined Pembrokeshire coast in Owen's day and the fish were
'very sweete and delicate meate and plentie taken'. Nevertheless with the
exception of north Pembrokeshire, Bardsey and Llŷn, lobster catching on
any scale was not practised until the last quarter of the nineteenth century
and in most parts of Wales it appears to have been restricted to 'gathering
by hand and hook around the rocks at low spring tides'.[9] After about
1880 there was a considerable increase in the number of boats fishing for
lobsters. Milford Haven, for example, had 14 lobster boats in 1892 and 25
in 1897. In the same year 10 lobster boats worked the north and west coasts
of Anglesey while two or three fished around Caldy Island from Tenby. In
Cardigan Bay north of the mouth of the Teifi and south of Pwllheli there
was hardly any lobster fishing although the rocky coast of the west was rich
in crabs and lobsters. Nevertheless local demand was low and those that
were caught were purchased by local hoteliers who were participating in a
growing tourist industry. Thus between 1902 and 1906, 631 lobsters and
120 crabs were landed at New Quay and Aberaeron and 70 lobsters and 83
crabs at Aberdyfi and Barmouth. These were small landings compared with
the 6,378 lobsters and 1,024 crabs landed at Milford and 4,090 lobsters
and 10,051 crabs landed on the Llŷn peninsula and Bardsey.[10] It seems that
until about 1918 lobster and crab potting was still largely concentrated off

Lobster fishermen, St David's Head, Pembrokeshire in the 1930s (Photo – M.L. Wight)

the Llŷn and Pembrokeshire coasts. Attempts were made in the inter-war period to develop lobster fishing in Cardigan Bay especially from the ports of Aberystwyth and New Quay, but quite often these fishing expeditions resulted in a heavy loss of gear so that lobster fishing in Cardigan Bay did not really develop until after 1945. Large motor boats, many of them converted ship's lifeboats, carrying perhaps 60 pots each were utilized and ports such as Barmouth developed rapidly into centres of lobster and crab fishing. With improved methods of transportation, appreciable quantities were taken by road and rail to English cities and throughout western Europe.

In addition to Cardigan Bay, the north and west coast of Anglesey also became important for lobster fishing and fishermen worked not only from' the larger ports such as Holyhead and Amlwch but also from small villages such as Llanrhuddlad (Church Bay) and Rhosneigr. In 1954/5 there were no fewer than 22 full-time fishermen and 16 part-time lobster potters operating from Anglesey coastal villages. In south Pembrokeshire 4 full-time and 4 part-time lobster fishermen operated from Tenby and Stackpole Quay. East of Tenby, despite the fact that lobsters are found along the rocky south Gower coast, potting never developed as an important fishing activity. A few were caught by hand and hook in the late nineteenth century, but they were all for local consumption. In 1982 however, small quantities of both crab and lobster from the Gower area were landed at Llanelli and Swansea.

After 1945, lobster fishing, usually carried out between April and September, was practised on all parts of the Welsh coast from Tenby in the south to Llandudno in the north. The central and northern parts of Cardigan Bay in particular developed as important lobster fishing areas. Not only did the number of lobster boats increase as herring fishing declined in such coastal settlements as New Quay, Aberystwyth, Barmouth and Pwllheli, but boats from such ports as Milford Haven and Fishguard and even from Breton ports paid regular visits to Cardigan Bay in search of both lobsters and crabs. Many of those were landed in the Cardigan Bay ports.

In Gwynedd in the post-war era, the main centres of lobster fishing were the rough north and west coasts of Anglesey and the Llŷn peninsula. Simpson points out that in Anglesey a total of 22 full-time fishermen and 16 part-timers operated in open boats from Anglesey ports such as Moelfre, Amlwch, Cemais and Holyhead. Most of them worked in open boats usually equipped with outboard engines. In Llŷn a total of 31 full-time fishermen operated in the summer months from such places as Aberdaron, Rhiw, Porth Ysgadan and Porth Oer while further south Barmouth, Aberdyfi and New Quay were important lobster-fishing ports. The rough grounds off St David's Head in Pembrokeshire provided a livelihood for a number of full-time fishermen operating from such creeks as Porth Glais, Abereiddi and Porth Stinan while four fishermen fished the southern coast of Pembrokeshire from Stackpole Quay and Tenby. According to Simpson a total of 118 full-time lobster fishermen operated in Wales in 1954/5. Today the number employed is somewhat similar to what it was in the 1950s although there has been a considerable improvement in both boats and equipment. Although small open boats manned by one or two men and carrying perhaps no more than half a dozen pots are still widely used, particularly by the increasing number of part-time fishermen, there has been a tendency in recent years to utilize purpose-built motor vessels carrying up to 100 pots each. At Aberaeron in Dyfed, for example, two steel lobster boats, each 40 feet long and broad beamed were used for lobster potting in 1984. Both boats were built at the Ynyslas boatyard in northern Ceredigion and each one had a wheelhouse well forward with a large open well for carrying pots aft. In many cases with the adoption of larger boats, fishermen are able to sail for considerable distances from their home ports, spend long periods at sea and are able to land their catches at ports other than their own. The appreciable quantities of lobsters landed at Barmouth, for example, within the last decade were mainly from the Milford Haven and Cardigan boats that harvested the lucrative fishing grounds of Sarn Badrig. Today substantial quantities of

both lobster and crab are landed in Wales and the Cardigan Bay lobster, found along rocky stretches of coast and also
in deeper water, is regarded as the most desirable type: the best weighing from 1½lb 2lb. Crabs are not considered as important as lobsters and some fishermen will even throw crabs back into the sea rather than bring them back to port.

Between 1979 and 1982 the following quantities of lobster and crab were landed at Welsh fishing ports.[17]

Crab and lobster landings (in tonnes) at Welsh ports 1979-1982

	1979		1980		1981		1982	
	Crabs	Lobsters	Crabs	Lobsters	Crabs	Lobsters	Crabs	Lobsters
Milford Haven	4.10	6.65	5.24	7.18	3.83		11.66	2.79
New Quay	0.21	5.01	0.42	0.93	2.07	5.85	1.69	5.10
Aberaeron	0.01	0.11	0.07	0.23	0.07	0.32	0.10	0.40
Barmouth	0.18	6.17	0.97	5.29	0.51	6.00	1.76	6.21
Aberystwyth	0.18	5.36	1.14	6.35	2.44	6.41	1.58	1.74
Aberdyfi		0.03	0.69	0.24	2.22	0.24	2.58	0.46
Porthmadog/ Criccieth		0.50						
Pwllheli	6.35	2.81	2.40	1.92	1.94	1.77	3.25	2.07
Aberdaron	2.49	2.42	1.66	2.07	1.26	1.25	1.06	3.85
Anglesey	0.23	2.72	1.08	2.36	0.90	1.99		0.73
St Davids (& Fishguard)			29.36	3.00	37.75	11.74	50.81	13.66
Conwy			0.10	0.20	1.37	0.26		
Swansea							0.15	0.10
Llanelli							0.67	0.29

Source: Welsh Office

The lobster pot

Although lobsters and crabs may be caught with hooks in shore-line crevices, all the commercial capture of the fish is undertaken with baited pots. The pot is a chamber containing the bait, the entrance to which is easy but the exit made difficult by a non-return valve, nearly always of the funnel pattern. The lobster pots used in Wales vary tremendously from region to region both in shape, size and materials. Willow, cane, wire, wood and plastic are all used to produce the best pot that a fisherman thinks can work effectively. Since the pots are made by each individual fisherman, many idiosyncrasies can appear in the final design of a pot. Thus, for example, one Cardigan Bay lobster fisherman in 1984 constructed a rectangular pot with a wooden framework, a wire body and a bottomless plastic child's sand bucket acting as an entrance to the pot. The whole was weighed down with a blob of cement at the bottom of the pot. Another fisherman on the west Anglesey coast had pots that consisted of rectangular metal bases each 20 inches by 17 inches consisting of 10 rods. A semicircular metal frame was welded to the base and covered with hemp netting: the only traditional component being a carefully woven basketweave entrance 8 inches in diameter and 7 inches deep. In the 1930s, Davis saw 'an ingenious type of French pot... at Beaumaris; it is cylindrical and the outer supporting hoops are the rims of bicycle wheels, joined longitudinally with broomsticks and the openings of the funnels are supported by the rims of small perambulator wheels'.[13]

Although there may be many variations in the detailed design of lobster pots from region to region and indeed from fisherman to fisherman, they may be broadly divided into three distinct types:

1. The so-called **Cornish ink-well** pot with the eye or eyes at the top. They are usually built of basketwork although plastic and wire ones have been

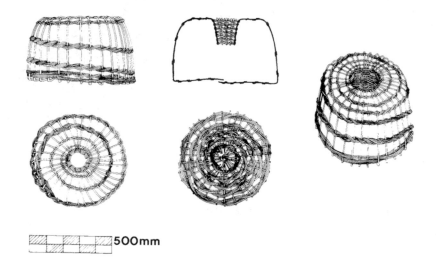

500mm

a) Cane lobster pot – Amlwch, Anglesey

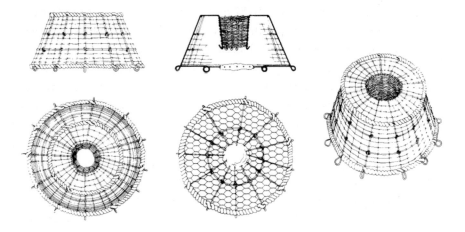

b) Inkwell lobster pot – St Davids Head, Pembrokeshire

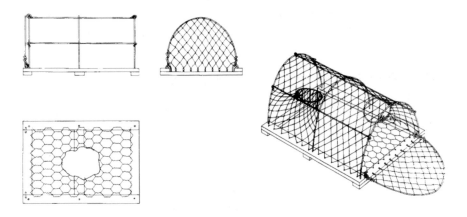

c) Double entrance, netted lobster pot – Barmouth, Meirionydd

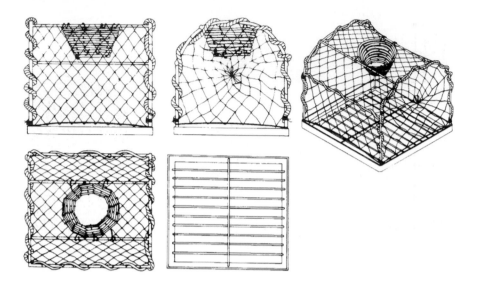

d) Single entrance, netted lobster pot – New Quay, Ceredigion

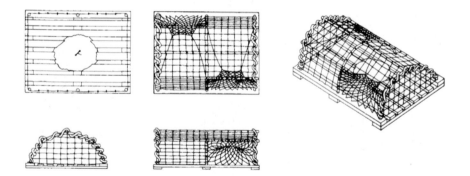

e) Double entrance, metal lobster pot – Llangrannog, Ceredigion

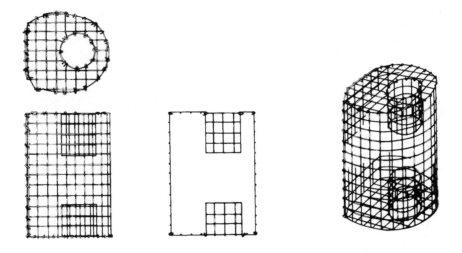

f) All metal double entrance cylindrical lobster pot – Milford Haven, Pembrokeshire

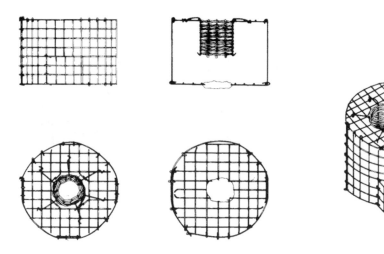

g) All metal inkwell lobster pot – Solva, Pembrokeshire

seen and they usually consist of ribs bound together with a spiral braid. As
the name implies, this type of pot originated in Cornwall and pots similar to
those used in south-west England are made by fishermen at such places as
St Davids and Porthglais in Dyfed from locally grown willow or imported
cane. They are flat-bottomed and are between 22 inches and 28 inches in
diameter and stand between 14 inches and 22 inches in height with a closely
woven basketworked entrance between 5 inches and 8 inches in diameter.
Pots as much as 5 feet in diameter were constructed as 'store pots' in which
lobsters were stored alive before being sent to market. They were of the
same type as those used for catching, but the mouth was covered with a disc
of wood or net to prevent the fish escaping.

A variation of the ink-well pot is found on the Llŷn peninsula and the
traditional pot of that region is a spherical willow structure with a single
entrance. The braiding of the pots is not of twisted withy as in south-west Wales,
but thin strips of unpeeled willow twisted continuously around the upright ribs.
Another variation in north and west Anglesey is an almost oval pot, again made
of willow. At New Quay in Ceredigion ink-well pots similar in dimensions to the
traditional Cornish pots are used but in this case the pot has a wire framework
covered with netting and weighted with a blob of cement poured into the pot
'and formed into a disc covering about half of the base and embedding the
wires, thus increasing the rigidity of the whole structure'.[14] Cubical pots of wire
netting on wooden frames with one or two entrances of wire or basketwork
that may be regarded as variations of the ink-well pot are widely used especially
on the west coast of Anglesey. These are heavily weighted with bricks or stones.

2. **The Scottish creel pot** usually consists of a rectangular base about 26
inches wide with a curved framework of wooden laths covered with wire,
hemp or nylon netting. The entrances – and there are usually two – may be
of 'soft' netting or wirework and the pot stands about 15 inches high. Creels
are usually fished in groups and although they are widely used in the north-
east of England and in Scotland, they were not introduced into Wales until
the mid-1950s when they were first brought to Mochras, Abersoch, Bardsey
and Aberystwyth. Although most creels are of net fitted over wooden half-
circle frames, metal frames covered with tarred wire netting have been used
in Llŷn and in general creels have become far more popular along the whole
length of the Welsh coast.

3. **Barrel-shaped pots** commonplace amongst Breton lobster fishermen
are used by some Welsh fishermen; the most usual type being of wooden
laths held in position by wooden hoops and weighted with stones. The ends
are closed by means of laths or netting with one or two entrances on the

side opposite the weights. Thus the entrances are always on the top of the pot when it is in use. In Milford Haven district a variant of the barrel pot is one made of galvanized iron rings held apart with wooden bars and covered with tarred netting, but on Bardsey a number of fishermen use the all-timber construction of the traditional French pot with a single entrance.

The extensive coastline of Pembrokeshire, both between Tenby, Caldy Island and Milford Haven and the coast north of the Haven in St Brides Bay and off St David's Head provide very rich grounds for lobsters and crabs. It is fully exposed to the south-west but the great potentialities of the coast have not been fully realized; indeed many of the lobster men of Milford fish the easier coast of Cardigan Bay. Nevertheless a number of full-time fishermen from such ports as Marloes and Dale fish St Brides Bay and a few from such coastal settlements as Solva, Porth Glais, Porth Stinan, Abereiddi and Porthgain venture into the rough seas around St David's Head and Ramsey Island. Their boats vary from 18 feet to 36 feet and the Cornish type of pot, usually of wire nowadays, baited with gurnard is used principally. Most of the boats are motor boats equipped with inboard engines and most are equipped with winches that act as pot haulers.

A considerable amount of lobster catching, much of it by part-timers from open boats, is carried out along the coast of Ceredigion and Meirionnydd, but undoubtedly of all the lobstering sections of the Welsh coast the most important historically and ethnographically are the Llŷn and Bardsey coasts. The lobster fishermen of such villages as Aberdaron, Porth Ysgadan, Porth Ger, Rhiw and Porth Meudwy as well as those from Bardsey itself were famed as highly skilled seamen operating in a particularly difficult coastal area. In that very rough section of the Welsh coast, the lobster was a very important item in the local economy and the fishing season that extends from April to September provided the region with one of its few cash crops. By late September, the activity has to cease for the rough winter weather makes it impossible to venture far from land and the severe gales witnessed in the area can, with the strong tidal currents, destroy a great deal of valuable equipment.

Unlike the lobster fishermen of south-west Wales, many of the Llŷn men clung to tradition not only using ink-well pots that have hardly changed in a century but also using wooden boats equipped with outboard engines that were built within the region itself. The *cwch gweithio* (working boat) of Llŷn was usually between 12 and 16 feet long with a broad beam that varied between 5 feet and as much as 7 feet 6 inches. Many of those used within the

Above and opposite: Eddie Williams, the Post Office, Aberdaron, Gwynedd constructing a traditonal Llŷn lobster pot, 1970

twentieth century were built to a traditional pattern by such notable boat-builders as John Thomas of Rhiw. The boats were usually built of larch, a very suitable timber that even if it cracked in the dry, would soon become waterproof when returned to the sea. Each of the small, broad-beamed boats was equipped with a single mast 22 feet high to carry a single sail. In 1947, John Thomas would charge £25 for a boat that was well-suited to the rough coastline of north-west Wales. Every fisherman would paint his boat either at the end or before the commencement of the fishing season and it was considered very unlucky to change the colour of a boat and thus risk a change in fortune.

The best lobsters are caught when 'the corn is starting to change its colour' and usually the boats leave the shore for the lobster grounds an hour before high tide to return again an hour or two hours after the high tide, but in the past a lobster potter expected to be out at sea for at least seven hours. Pots are laid singly or strung out in groups of twenty or thirty and fishermen believe that if shoals of pollack are present along a section of coast, then that is a good omen for a successful lobster catch.

Many Llŷn lobstermen are employed full-time in potting during the season but in the past many combined fishing with such tasks as rabbit catching, general labouring and of course preparing boat and equipment for the fishing season. The lobster fishermen or *cwillwrs* were certainly highly skilled men who operated in very rough seas with many dangerous currents and they were equally skilled in the task of preparing the unique, spherical lobster pots traditionally used in north-west Wales. The traditional Llŷn pot is of willow, measuring about 18 inches in diameter with a single woven entrance (*y cnyw*) 7 inches or 8 inches in diameter. In the past, winter-cut, two-year-old willow was utilized for lobster pots. This was obtained from withy plantations in Nefyn, Pwllheli, Afonwen or Abersoch and was purchased from farmers, in the 1950s for half-a-crown a bundle. The

Casting a lobster pot, New Quay

bundles were left to season, perhaps under sacks, for about a month before they could be used. The bark was left on the willow in the finished pot for many an old fisherman maintained that the dark colour of the bark would make the spherical pot invisible to the foraging lobster. Many old fishermen too, refused to adopt cane pots that became popular after about 1948 because they were far too conspicuous in the water. Others maintained that cane was far better because it was much smoother and fish would slip far more easily into the baited trap. But whether the pot is of cane or of willow, its design and method of construction is similar, except that where willow is used each rod is split into two before weaving, whereas cane is left in the round.

To construct a pot, a task usually performed in the winter months, a round disc of wood perforated with about a dozen holes is taken to begin the entrance to the pot. A rod is inserted into each hole and the funnel-like basketwork entrance is constructed. As the work on the *cnyw* is completed, longer rods that form the ribs (*asennau,* sing. *asen*) are inserted. These are fairly stout pieces of withy and the longer ones, 36 inches in length, are inserted first and these are followed by shorter rods 32 inches and 24 inches long and all woven to a spherical shape. Thinner, more pliable rods (*eiliad* or *y bleth*) are then braided in a continuous line around each rib to provide a very strong, round basket. The bottom of the pot is finished carefully by weaving all the rods together. With the sphere completed, a piece of string (*pont,* pl. *pontia*) to carry the bait is tied to the *cnyw* and potbottom and a piece of rope (*y sgop*) with cork markers attached is tied on. A cork is inserted every yard or *gwrhyd* along the *sgop* and this piece of rope may be as long as 20 yards. In any case it has to be considerably longer than the depth of water for which the pot is designed due to the strong movement of current and tide that may easily move anchored pots into deeper water. Dogfish, gurnard, conger eel, mackerel and even rabbit meat are all used as bait. It is believed that crabs prefer fresh bait but the lobster prefers bait that is two or three days old. Each lobster pot has to be weighted down with three stones attached to the outside to produce a total weight of as much as 25lb. per pot. Each rope used to attach the stone weight is known as a *llyfeithar* or *carcharor*. Of course it is vital that the catch of lobsters and crabs be kept alive before being collected by the fish-merchants and every fisherman has a wooden keeping creel (*cawell gadw*) anchored close to the shore for this purpose, described here by an Aberdaron man:

> Our keeping creel was 3 feet long by 2 feet wide and was no more than 9 inches deep. We tried to put a padlock on our keeping creel to prevent the contents being stolen, but this was not successful. Some *cewyll cadw* were rectangular boxes with holes in them but the best were exactly like the catching pots, but without the woven entrance. We closed the entrance with a round piece of wood and a lobster could be kept in the creel for about a week without being fed at all.[15]

enacted but this was almost too late to ensure the continuity of the beds. A half-hearted attempt was made to transfer spat from Stackpole but gradually the industry declined as the beds became worked out.

In Tenby, as elsewhere, oysters were caught in dredges hauled along the sea bottom from a boat. A dredge consisted of an iron frame, some 2 feet square, and it had a sharp knife blade or rake that actually scraped the sea bed. This blade was inclined downwards and forwards to facilitate the scraping action. A net bag of strong twine at the top and of leather or iron rings at the bottom was attached to the metal frame in order to catch the oysters as they were scraped from the sea bed. When dredging, the boats drifted or sailed slowly over the oyster beds, the heavy dredge scraping the sea bottom to catch the fish in the bag.

In Tenby the open boats, up to 25 feet in length, were locally built luggers all constructed to a similar design and rig. In 1864, between twenty and thirty luggers operated from Tenby harbour, by 1891 the number had increased to forty-nine. Tenby luggers were usually clinker built but later ones were carvel built. All had full mid-ship sections and were on average 22 feet in overall length, but tended to be built larger as older boats were replaced.[11] The forepart was decked to the mainmast and underneath was a small cuddy. There were usually three rowing thwarts and a thwart and quarter benches in the stern for the helmsman. Since the lugger was not hauled out of the water except for maintenance work, they were usually built with a heavy keel, usually 18 inches deep. Heavy timbers of oak were used for the floor and were alternated with bent frames. The vessel was equipped with two masts; a tall mainmast stepped about one third of the boat's length from the sternpost and a mizzen mast, about half the height of the mainmast. This was stepped in a socket on the transom and followed the rake of the transom. When a bowsprit was carried it was a long spar that passed through an iron ring to the starboard of the sternpost with the heel secured in a socket on the foredeck near the mainmast. The light outrigger spar for the mizzen mast was usually of ash and was stepped through a hole in the transom. Although normally the mainsail was a dipping lug with two or three rows of reef points, for oyster dredging this was replaced with a try-sail held to the mast with hoops. Three jibs of various sizes were set appropriate to the strength of the wind and a small spritsail was set on the mizzen.

By the late 1940s, the Tenby lugger, used not only for oyster dredging but also for other kinds of fishing and for pleasure trips for summer visitors, had virtually disappeared from the sheltered Tenby harbour.

By far the most flourishing of all Welsh oyster fisheries was that of Swansea Bay where at Oystermouth and Mumbles the oyster fishery in the

1880s still 'employed about four hundred men during eight months of the year'.[12] In the seventeenth century, the oyster fishery was said to be the best in Britain but by the last quarter of the nineteenth century the stocks had become considerably depleted due to over-fishing and lack of conservation. 'During this period', says Lloyd,

> millions of oysters were destroyed without regard to age or fitness while thousands more were taken away to restore the beds in the east of England. The consequences were inevitable; the yield began to decline and by 1874 or 1875 the fishery was becoming impoverished and there were more skiffs than it could support. From about 90 skiffs in 1863 and roughly 68 in 1866 the number had risen to 188 in 1871 ... seven years later the number had fallen to 47 skiffs.[13]

A measure of the decline can also be obtained from the figures given for the number of oysters taken during this period which amounted to 9,050,000 in 1873, 6,600,000 in 1874 and 3,810,000 in 1875. By 1894 only 600,000 were taken in the year. By the 1930s the once flourishing industry that employed so many people in Oystermouth and Mumbles was completely dead. In the heyday of oyster fishing in Swansea Bay the fishermen operated in skiffs arid in half-decked boats. Between high and low tide marks in the Oystermouth district were a series of spaces or 'perches' marked out with tarred stones and each fisherman had his own perch.[14] He paid a small rent to the duke of Beaufort, the owner of the foreshore. The oysters were dredged and the catch emptied into the perches to be kept alive until required for market. Each perch was marked by the buoy of the owner. In Swansea Bay, too, special areas were laid aside as nurseries for immature oysters and these were known as 'plantations'. The plantations were generally farther out to sea than the perches and were uncovered only at low-water spring tides.[15]

The principal oyster fishery village was Oystermouth at the most westerly end of Swansea Bay. Mumbles Head protects the village from the south-westerly winds and at high tide the sea reaches up the shingle beach to within a few yards of the road. At low tide the muddy bottom is well suited for grounding boats. In 1914 the village was described as 'the chief station for oyster fishing and the fisher folk there devote themselves almost exclusively to that business ... The Mumbles dredging fleet consists of about twenty sailing smacks carrying three hands apiece'.[16] That figure of only 20 dredging vessels showed a great decline in the industry from its heyday in the mid-nineteenth century. In 1864 it was estimated that the number of decked sailing vessels operating from the Mumbles-Oystermouth area was 66; 8 vessels worked from Swansea and a further 22 from Port Eynon.[17] In the 1860s the area was obviously being overfished, and in addition spawn was being transferred in considerable quantities to the Whitstable district of Kent. Fishermen from other areas, from Colchester in Essex and from Jersey, invaded Swansea Bay and it is said that 'no fewer than 200 skiffs and 1,000 persons were engaged in the trade'.[18] Although this was an over-estimate, it is obvious that Swansea Bay was being overfished.

Until about 1860 when Appledore-built, decked vessels replaced the old, locally built open boats, the fishermen limited their activity to dredging the oyster beds in Swansea Bay and off Mumbles Head that were close to the shore.[19] With the widespread adoption of large-decked vessels they could wander much further afield. Some were based in Tenby and dredged the 'Abyssinia Haul' off St Gowans Head. Others dredged the 'Metz Haul' off Porthcawl, the 'Roads Haul' off Mumbles Head; 'the Green Grounds' and 'Jersey Haul' in Swansea Bay or beds off the Helwick shoals. 'All the oyster patches had local names such as "Bantum", "Boggy Hole", "the Black Ones", "the White ones", [probably the white oyster ledge off Mumbles Head] and they were located by picking up shore marks. Few skiffs if any, carried compasses and they must have found it extremely difficult to locate the dredging grounds in thick weather.'[20] Usually

Swansea Bay fishery. General view looking towards Mumbles Head

a decked skiff with a crew of three operated two oyster dredges, each one weighing about 100 lb. and after sailing first to windward and then down wind over a patch a dredge could weigh anything up to half a ton. A mast winch or even a steam capstan was utilized to haul in the catch which was emptied on the deck and then sorted. Waste material was shovelled overboard and the oysters put into a hold. Anything up to 3,000 or 4,000 oysters could be caught in a day, although one optimistic estimate said that 8,000 could be caught on a good day.[21] The proceeds were divided into four – one share for each member of the crew and one for the boat.

The oyster fishing industry was not limited to the villages of Mumbles and Oystermouth for the south Gower village of Port Eynon was also an important centre. The 'Bantum patches' at the eastern end of the Helwick shoal was their main dredging area and the season extended from September to March. Boats were manned by four crewmen each and in the 1840s it is said that 40 skiffs were in regular use in Port Eynon. The industry finally closed in 1879.

One of the main factors that contributed to the demise of the oyster fishery in Swansea Bay was overfishing for 'the goose that laid the golden egg was killed and now [in 1899] some dozen boats are all that engage in the fishery'. In desperation a number of the Mumbles fishermen made the

Oystermouth harbour – the centre of operation for the Swansea Bay Fishery

long voyage to the Solway Firth in order to dredge oysters there and they would remain in the Firth for up to six weeks before returning with their catches to Mumbles. They never landed at another port. Pollution of the beds by industrial spillage into Swansea Bay also had an effect on the oyster beds. Between 1918 and 1920 disease hit the beds and they never recovered and by 1925 the activity had ceased. The beaches from which the skiffs operated had no breakwaters and no jetties and by 1930, only three vessels, the sailing skiff *Emmeline*, the motorized skiffs *Secret* and *Rising Sun* were all that remained of the once extensive fleet of Mumbles oyster skiffs.

The oysters from Swansea Bay were taken by boat but later by railway to the London market and some were sold locally. D.M. Jenkins, a native of Oystermouth, said in an interview in 1969 'My father was an oyster dredger and it was a custom to bag and send oysters which the children used to bag and send them by the old Mumbles train. They used to be sold at four shillings for 120 oysters. My father also had a table on the road to Mumbles Pier and it was a custom for people to stand and wait for them to be opened at seven for sixpence, pepper and vinegar included.'

Until the mid-nineteenth century the oyster fishermen of Mumbles and Oystermouth used open boats, usually equipped with two masts, but around 1850 in order to obtain boats that could go further afield a number of Mumbles fishermen went to Colchester in Essex to see the boats used there. As a result a similar type of vessel to the Essex smack was adopted by the Welsh oystermen and this type of single-masted skiff soon replaced the locally built open boats that had been used by generations of fishermen. Most of the new vessels that came in after about 1858 were built at Appledore and at other yards on the north Devon and Cornish coasts and had to be taken back there for major repairs. 'But all the maintenance and smaller repairs were carried out at Mundick yard which stretched along the foreshore opposite the Mermaid Hotel in Mumbles. It was in use from about 1860 until 1892.'[22] Most of the skiffs were from 37 feet to 40 feet in hull length with beams of between 10 feet and 11 feet and between 5 feet and 6 feet in depth. They had low freeboards aft in order to facilitate the hauling in of the heavy dredges and had a hold aft of the single mast. The mainmast was fairly short, being usually about 70 feet long with a light topmast, about 18 feet long above it. The vessels were equipped with a mainsail and staysail above and a working jib of lighter canvas.

In north Wales, oyster fishing was far less important than in Swansea Bay and Milford Haven, but oyster fishing was carried out in the Menai Straits and at Rhoscolyn and Beaumaris in Anglesey. It was also pursued in Caernarfon Bay where French fishermen worked with some success in

The Mumbles oyster skiff Emmeline *built in 1865 at St Ives, Cornwall for William Burt of Oystermouth*
(Photo – R.J.H. Lloyd)

The Mumbles oyster skiffs Snake *and* Hawk *(Photo – R.J.H. Lloyd)*

the mid-nineteenth century. Below the city of Bangor the so-called *Gored y Git*, originally a salmon weir, was converted in 1852 for oyster breeding.[23] In 1908 records state that 19,503 oysters were taken in the Menai Straits, but after that date there is no evidence to suggest that any more were dredged. In west Wales, between Newport and Cardigan an oyster bed, five fathoms deep, that extends from three to four miles in length is regarded as one of the richest in Wales, but this has never been utilized.[24]

Despite many attempts at reintroducing oyster fish into Wales during the twentieth century, the industry has declined greatly. In 1986 an oyster fishery was established in Milford Haven around the village of Lawrenny dredging for oysters in the waters of the Haven. This has proved to be a profitable and successful venture and Milford oysters are now widely exported. Oyster beds in Swansea Bay promise to become fruitful by the mid-1990s.

Scallops

The scallop (*Pecten maximus*) and queen scallop (*Chlamys opercularis*) – close relatives of the oyster –were caught in great quantities in Cardigan Bay until the 1980s. New Quay in Ceredigion was particularly important for the capture of scallops and a processing plant on the foreshore employed a number of inhabitants of the village until 1985. Most of the shellfish, in common with the lobster catch, were exported to the European Continent but overfishing especially by Spanish and French fishermen led to the sharp decline in catches and to the subsequent closure of the New Quay processing plant.

The scallop or clam, a bivalve shellfish, occurs widely along the western coast of Wales and is commonest in depths of 10 to 20 fathoms. The shell of the scallop bears concentric rings, one of which is laid down each year as the scallop grows. At four or five years old the fish is at its best. It occurs mainly on bottoms of sand or sandy or muddy gravel; conditions in parts of Cardigan Bay are its ideal habitat. The closely related but smaller queen also lives in a similar sea bottom and is commonly caught with the scallops. Unlike the scallop which measures 5 inches long on average, queens were rarely sold in the past, but their flesh was widely utilized as bait in long lining for mackerel, cod and other fish. Their chief drawback is the fact that they require far more effort in shelling than the true scallops in order to obtain the same amount of meat. In addition, transport costs of shells was prohibitive. Nevertheless, due to the abundance of the fish and the opening of a processing plant around 1960 it became an economic proposition to catch and process queens. The meat was frozen at the processing plant for export. In the late 1970s new queen beds were exploited in Cardigan Bay both by local fishermen and by French and Spanish boats who paid regular visits to the Bay. Scallop fishing is traditionally a winter and spring occupation, the roe being at its best in early spring before spawning. Nevertheless by late spring the roe recovers rapidly and fishing can be renewed almost immediately. The fact that a processing plant was established on the Ceredigion coast overcame the difficulty of transporting scallops to the Billingsgate market in warm, summer weather so that scallop fishing was, in recent years, practised virtually throughout the year even when no 'r' appeared in the name of the month. Deep freezing contributed substantially to the scallop-dredging industry.

The equipment used for the capture of scallops and queens is somewhat similar to that used for oyster dredging. The standard dredge consists of a triangular wooden frame with a toothed crossbar set at an acute angle to scrape the scallops from the sea bed. The scallops are caught in a large bag attached to the frame, the belly of the bag being usually of steel rings that can withstand the chafing along the sea bottom and the back of stout twine netting. The mesh of the net is such that it allows small stones and gravel to pass through, retaining the fish in the bag. The width of the dredge vanes from 4 feet to 6 feet and a boat may tow two or more dredges from a bar attached by a bridle to a single warp. Motor boats of 25 feet or more in length are used exclusively for scallop dredging. In fishing 'the dredge is thrown over the stern the right way up and the boat travels at full speed until an amount of warp is paid out equal to approximately three times the depth of water. The dredges are towed slowly for one to one and half hours and then hauled aboard by a power driven winch. The catch, consisting of scallops, mixed with stones, empty shells, sea urchins, starfish and other objects is sorted on board.'[25] The standard scallop dredge is not efficient for the capture of queens, for they are far more mobile than the scallops. With the approach of the dredge many queens escape by clapping the valves of the shell together and swimming away from the dredge. Trawling with beam and other trawls is a far more effective way of catching them.

Within recent years the gathering of scallops by skin-divers has developed off the coast of south-west Wales. Divers are able to operate in scallop beds located in rocky or stony patches where it would be impossible to use a dredge. Only the largest scallops are gathered in this way and there is a severe limit on the depth of operation.

The number of scallop fishermen from Brittany and Spain in particular have increased greatly and dozens of boats may occupy the Ceredigion and Pembrokeshire coast on any day. This has caused a furore amongst native fishermen who fear over-fishing and depletion of fish.

7 NETTING, TRAPPING AND LONG LINING

Historically, the herring was by far the most important fish caught off the coast of Wales and the history of herring fishing in the Principality goes back for many centuries. Other varieties of fish were, of course, caught in considerable quantities in Welsh waters and as the herring stocks became exhausted, other fish assumed greater commercial importance. Some of the other main species were only of relatively short-lived importance; the stock was soon depleted as the once prolific grounds were overfished by both native and foreign boats. The best example of this was the hake that once occurred in great quantity in the Irish Sea and off the south coast of Ireland. Until the end of the nineteenth century, the hake had little commercial significance, but after about 1890 it became the commonest whitefish landed in Welsh ports as the fishing grounds in the Irish Sea and beyond were exploited by the new steam trawlers that were able to travel to the more distant fishing grounds and spend considerable periods there before returning to port.

The growth of Milford Haven as one of the great fishing ports of Britain and to a lesser extent the development of Swansea and Cardiff as fishing ports, was tied to the commercial exploitation of the rich hake fishery of the Western Approaches. Stupendous quantities of large hake were landed at these three ports before the 1914-18 war and that variety of fish was by far the most valuable landed. At Milford in 1913, for example, no less than 10,559 tons of hake were landed while 4,575 tons were brought to Swansea and a further 3,818 tons to Cardiff. Over 90 per cent of all the hake landed in the United Kingdom came to those three ports. But the prolific grounds were overfished and the first indication that they were being destroyed was the capture of much smaller and younger fish by the trawler-men after 1918. By the late 1920s, the stocks were greatly depleted so that in 1928 hake landings at Milford for the first six months of that year were less by 33,000 hundredweight than they were for the corresponding period in 1925.[1] The trawlers had to go further afield, to the Porcupine Bank off the west coast of Ireland for example, in search of hake but catches had declined to a very low level by 1939. The stocks were replenished during the idle years of the war but after 1945 intense fishing contributed to the complete destruction of the hake grounds off the Irish coast. In 1961 no more than 9,162 hundredweight of hake, mostly from distant fishing grounds, were landed at Milford. Today, the hake that was of such great commercial importance in the formative years of the Welsh trawling industry is no longer of any significance. In 1983, only 30 tonnes of hake was landed in Welsh ports. Nevertheless, as hake catches declined, other fish particularly mackerel and varieties of the skate and ray family have increased greatly. In 1983, for example, 665 tonnes of skate were landed in Wales of which 435 tonnes were landed at Milford. For the same year 2,813.68 tonnes of mackerel were landed in Welsh ports with 2,780.15 tonnes being landed at Milford alone. Much of that vast catch was destined for a West African market and was in itself a decline in the total catch for previous years. In 1979, 47,760.70 tonnes were caught off the Welsh coast; in 1980, 35,501 tonnes, in 1981, 12,007 tonnes and in 1982, 5,648.50 tonnes. Perhaps this is a repeat of the story of the herring and the hake with overfishing contributing to the virtual extinction of an important variety of food-fish.

The seas around Wales are well-blessed with a wide range of underwater creatures. There are demersal or bottom-feeding fish such as soles, plaice and cod, and pelagic fish such as herring, sprat and mackerel that occur nearer the surface of the water. The techniques of catching those two main groups differ considerably for while drift nets, free to move with the wind

On board a Milford trawler, 1923

and tide are used for the capture of 'swimming fish', the trawl is the most effective instrument for the capture of those fish that live on or near the sea bottom. George Owen of Henllys in his *Description of Pembrokeshire* of 1603[2] describes the great abundance of pilchards and mackerel off the coast of south-west Wales, 'but of these two sortes nothinge in respect of the hearinge' that was the all-important fish in the seventeenth century. Turbot and brill, sole and plaice, ling and cod were all caught but as Owen pointed out 'yt would require a particular volume to write of everye severall kinde'. Undoubtedly the Pembrokeshire coast was one of the richest fishing regions in the whole of Britain and ports such as Milford Haven and Tenby developed into major fishing ports. That coast, together with the Bristol Channel was important enough to attract 'the Brixham smacks [that] used to come regularly to the Channel, from eighty to a hundred of them, working from Milford every summer before the [first] war.'[3] Herring, mackerel, hake, cod, ling, whiting, pollack, bream, conger eels, skate, and sole were found in great profusion. The sole in particular was very abundant in the Bristol Channel and 'Barry soles' were famous for their delicacy of flavour.[4] The halibut and haddock, essentially 'northern' fish were rare in the Bristol Channel although quantities of both species from Irish Sea fishing grounds were landed at Milford and Swansea.

Just as the Bristol Channel and the south-west peninsula of Wales are open to the Atlantic, so too is the coast of Anglesey in the north well-placed in relationship to the movement of migratory fish. Between November and January, for example, extensive shoals of plaice gather in Red Wharf Bay before migrating to the spawning grounds of Cardigan Bay. In that area they are found close inshore in late winter and early spring, but later in the year they move into deeper and cooler waters. Other flatfish such as turbot, brill, flounder, sole and dab occur widely near the coast of Gwynedd and Clwyd and varieties of roundfish such as pollack, cod, whiting, mackerel, mullet and ling are caught in appreciable quantities. In 1983, for example, the principal fish landed at Conwy was plaice (52.44 tonnes) followed by skate (34.24 tonnes) and whiting (15.43 tonnes). Small quantities of hake, saithe, sole, dogfish, monkfish and turbot were landed, the total of whitefish landed being 130.36 tonnes. At Holyhead in 1983 the principal fish landed was dogfish (194.37 tonnes) followed by cod (29.25 tonnes) and whiting (18.28 tonnes). A total of 251.32 tonnes of whitefish was brought ashore at the port. Caernarfon and Bangor are also fishing ports and a total of 295.96 tonnes of whitefish was landed at these Menai Straits ports in 1983. In this case however, the plaice (77.70 tonnes) and skate (77.50 tonnes) were by far the most important.

The waters of Cardigan Bay are relatively shallow and likely to be greatly affected by disturbance from the Atlantic and by strong tidal currents. Migratory fish such as the herring and mackerel were caught in large numbers in the past, but fishing for whitefish never developed to the extent that it should. Fishing ports were always small; many of the fishermen were part-time workers and it was very difficult indeed to send fresh fish to a lucrative urban market. Bass, mullet, bream, whiting and other species occur in considerable numbers, but fishing ports such as Aberystwyth, Aberdyfi, Aberaeron and New Quay were not of great significance in the commercial development of the whitefish catch in Wales. In 1983, for example, only 8.22 tonnes were landed at Aberystwyth, 42.40 tonnes at New Quay and 6.78 tonnes at Barmouth. The principal fish landed was mackerel followed by skate with minor quantities of cod, plaice, sole and turbot.

There has been a spectacular decline in the fishing industry in Wales within the last fifty years or so; a period that has seen the virtual extinction of some fishing fleets that were once flourishing. Cardiff, for example, in 1900 had as many as 13 trawlers of considerable size but is no longer concerned with the fishing industry; Swansea that once possessed an extensive fleet is less important as a fishing port than Conwy or Bangor. Milford Haven, once ranked as the fourth fishing port of the United Kingdom, saw the demise of its trawler fleet in 1991 whereas in 1908, 44,000 tons of fish were landed at the port. Even in 1932 when the fishing industry was already on the wane as a result of overfishing in the Irish Sea, as many as 108 fishing vessels were based on the port and 150 others visited Milford for eight or nine months of the year.[5] No longer does the daily fish train which left 'Milford for all parts' run at 3 p.m.; the old fish market was demolished in the spring of 1991; and the golden era of the port that lasted no more than sixty years is definitely over.

Throughout Wales the story of decline has been the same, for the fishing industry once so important in the economic life of coastal communities is today of little consequence. With the rapid decrease in the number of merchant seamen usually employed in inshore fishing when they were home on leave, manpower has declined. 'Nothing can be done without men' said the 1944 *Report of the Superintendent of the Lancashire and Western Sea Fisheries District*. 'The type of men that have usually carried on inshore fishing are gradually dying out and are not being replaced with younger men ... In the past it was mostly merchant navy men that were attracted.' That merchant navy no longer attracts the men from the coastal villages of Wales as it did for centuries. In the spring of 1988, for example, the total number of fishing boats operating in Wales amounted to 153, most of them under 18 feet in length working on a full-time basis, and 297 on part-time work. These employed 224 full-time fishermen and 383 part-timers. Of course, in the winter the number employed was considerably lower and for the fourth quarter the total number of boats employed in fishing amounted to 85 full-time and 143 on part-time work. Winter fishing employs no more than 135 full-time fishermen and 223 part-timers.

Fixed nets

A method of fishing that has lost much of its popularity amongst inshore fishermen in recent years is the fixed net. A large range of fixed nets for capturing a selection of fish from shrimp to cod were once widely used in the sea near Cardiff, in Swansea and Carmarthen Bays and along the north coast of Gwynedd and Clwyd. The nets were used mainly on relatively flat beaches with a considerable tidal range. On the mud flats to the east of Cardiff, for example, where low water could reveal flats as much as three miles wide, fixed nets were utilized for the capture of mullet, skate, whiting, sole and even the occasional salmon. The nets used were simple walls of netting about 300 yards long and about 3 feet deep tied on to a row of posts. The posts were usually about 6 feet high and set about 10 feet or 12 feet apart and they extended to within about 18 inches off the ground. The nets were known as 'hang nets'. Until about 1940, the stronghold of stake-netting was Swansea Bay, between Swansea itself and Mumbles Head:

> Here are several large V-shaped nets locally known as 'Stop Nets' or 'Kettle Nets'. They are the largest nets of the type found in the country, the total length of the two arms of each net amounting to about 700 yards. They are usually set in series with the tips of the arms of adjacent nets almost meeting and with the axis of the net almost at right angles to the shore, the tips of the arms not reaching high water mark. The net is about 7 feet high set on stakes and of 1 inch bar mesh, the apex there is a circular, roofed cage about 12 yards in circumference, prolongations of the arms into which form the usual type of non-return trap. Local by-laws require the cage to be in such a position that a pool is left at low water.[6]

Until 1939, stake-nets were far more common in river estuaries and along the Welsh coast, than they are today. Some of the nets, such as those used in Carmarthen Bay, were simple gill-nets, 4 or 5 feet high and never more than 100 yards long, fixed to stakes and allowed to swim with the tide. The fish stopped by the net would be evenly distributed along the whole length of the set and would not be guided to one section as in the Swansea 'Kettle Net'. Davis describes the so-called 'Ferryside stop net' as being '100 yards long and 14 meshes deep, set in by the half. In some cases the end is twisted round ... forming the "bung end" which functions like the in-turning arms of fish weirs, of which it is an exaggerated form.'[7] Undoubtedly this has always been a characteristic of Carmarthen Bay stake-nets, for an observer in 1863[8] states that most of the nets between Carmarthen and Llanelli were shaped like 'a ram's horn'.

A slightly more advanced method of setting stake-nets was to place them in zigzag fashion, to form a series of Vs, as at Llanfairfechan and Aber on the north Wales coast. Each arm of the V was about 50 yards long, but considerable variation in length could occur, according to the configuration of the coast at a particular point. In the Dee estuary, the stake-nets were allowed to rise and fall freely with the tide. These nets could be simply single walls of gill-netting or they could be fixed trammel-nets. In the latter case, the nets that were widely used in the Connah's Quay district on the Dee consisted of three walls of netting -lint and armouring set across a width of river. The centre wall was of small mesh and loosely hung on the head-rope and attached to a lead line; the outer walls of coarser mesh were set tightly on the head and lead lines.

By far the most complex of stake-nets, widely used on the north Wales coast, from the mouth of the Conwy to the Dee, was the so-called 'bag net' introduced into the area by Scottish fishermen in the 1820s. In the Conwy estuary, for example, the Salmon Commissioners' Report of 1861 notes that 'about twenty-five years ago a party came [to the Rhyl district] from the Solway Firth and put some stake nets here, and they took a great deal of salmon that way ... The stake nets were cut one night by some ill-disposed persons.'[9] In the Conwy the so-called 'Scotch weirs' were introduced by Scotsmen in about 1820 and were regarded as a serious obstruction to navigation both on the Conwy and the Dee.[10] The Scotch stell operated,

upon the principle of a leader running from or near high water mark seaward, against which the salmon, in their course along the coast, strike and, in their endeavour to find a passage, are guided into a narrow opening, the entrance to a chamber or trap, from which there is no escape. In some cases, these nets are of great extent, and have many chambers, the last being placed so far into the sea or channel as the very lowest tides will permit; it is never entirely dry and is generally waded into and fished with a scoup net, and to this chamber what is called a bag or fly net has of late been attached, which reaches still farther seaward ... It appears they were first introduced on the Scottish side of the Solway Firth about the year 1780.[11]

The introduction of Scotch stells was objected to violently by the fishermen of north-east Wales. Those in the Clwyd and Dee estuaries were destroyed by local people in the 1850s and they only remained on the Conwy, where they were leased by the Corporation of the town to some immigrant fishermen. By 1870 they had been largely replaced by seine-nets.

In most of the places where stake-nets were used, the sandy beaches were firm enough for the fisherman to walk to his net to collect the catch without the necessity of using a boat or other device. In other sections of the coast, however, the muddy and soft nature of the foreshore made it virtually impossible for the fisherman to walk to his nets without some kind of device to assist him in his journeys. The most notable of these was the unique 'mud sledge' or 'mud horse' used by fishermen for reaching the hang-nets that were once commonplace on the vast expense of mudflats east of Cardiff.[12] This device was also known on the north Somerset coast and was indeed still in use near Hinkley Point as recently as 2008.[13] It may indeed have been introduced into the Cardiff area from the west of England. It was used in Cardiff until the early 1930s for travelling to the hang-nets that could be as far as three miles away from the coast. With the huge tidal range of the Bristol Channel, it was essential to move quickly to the nets and return again before the rapidly incoming tide obliterated the mud flats. The mud horse consists of an elm board shaped like a surf-board measuring from 6 feet 6 inches to 7 feet 6 inches and about 18 inches wide. The board acts as a runner and the upturned front is kept in position by a rope attached to the front and to the rectangular wooden framework built on the runners. That framework, usually of pine, measured from 36 inches to 48 inches high, depending on the height of the fishermen that worked it. 'It is essential that the height of the bar is correct for the individual using the horse; if too high he will not be able to rest his weight on it; if too low it will press uncomfortably into his stomach and the movement of his legs will be cramped.'[14] To use the horse the fisherman leans over the top bar and pushes with his feet on the surface of the mud and it is said that a skilled user can travel over the mud 'as fast as a horse can canter over firm ground, so that the appliance is well suited for its purpose'.[15] On each horse a canvas or net bag is nailed to the top surface of the rectangular framework and the catch is carried in this.

Another type of net used in the Cardiff district was the so-called 'Mallinger Net' used for the capture of shrimps and sprats. This was a single net in the form of a bag, 7 feet long and 24 feet high. The bottoms of the fixing poles were buried in the sand and the foot rope of the net was set about 10 feet from the bottom. The net moved freely with the tide for the capture of the small fish that were collected from a boat at about four hours ebb.

Fish weirs

Undoubtedly the trapping of fish is one of the oldest methods of fishing known to man and throughout the country, in rivers and along the coast, may be seen the remains of stone or wattle weirs, many of them erected centuries ago. The widespread occurrence of the word gored as well as other names associated with the artificial damming of the flow of water, give some indication of the importance of weirs in the economic history of rural Wales. The two most common names for weirs are *argae* and *cored*.[16] *Argae* does not always refer to fishing weirs for it could refer to a weir built in association with a mill, but the word *cored* on the other hand refers invariably to fishing weirs, especially those constructed of plaited withies.

Some fishing weirs were constructed in such a way as to completely stop the mouth of a narrow tidal creek. On the ebb, water would run off, leaving the fish stranded behind the barrier of stone or brushwood. 'As soon as it was found that this barrier blocked the ingress of fish' says Davis[17] 'some arrangement would be made to allow them to pass the barrier while the tide was flowing. This effect would be produced by an opening in the wall which would be closed at high water.' Examples of this type of weir survived until the 1920s in the 'fish ponds' of south Devon.

The most common type of weir was that especially designed and constructed to fish along the coastline:

> The simplest barrier of this sort would be V-shaped, with its apex pointing in the direction of the ebb-tide, so that all the fish coming inshore with the flood tide can easily pass the wedge-shaped obstruction, but with their return with the ebb, come between the arms and on the fall of the tide are gradually stranded in the apex ... They are generally built of stone and stakes or hedging, each one being anything up to 400 yards in length and the apex being near low water mark.[18]

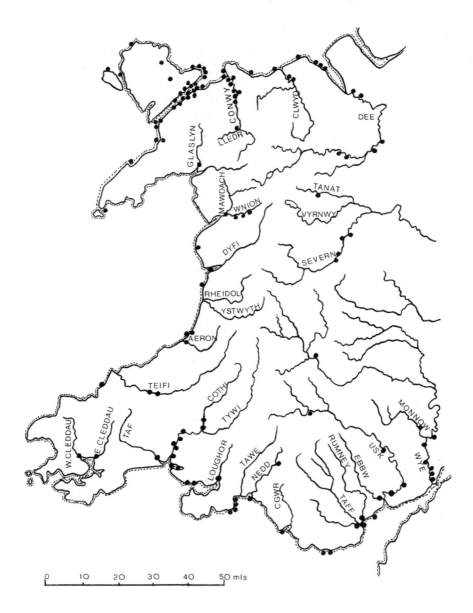

Fishing weirs in Wales

The Menai Straits and the shores of Cardigan Bay in particular had many fish weirs. In the Llanddewi-Aber-arth district of central Cardiganshire, for example, it is said that there were about a dozen fish weirs operating in 1861;[19] by 1896 these had declined to nine and in 1924 only two weirs remained in operation. All these were concentrated between the mouth of the Aeron and the mouth of the Arth, and they originally belonged to the monks of Strata Florida:

> The making of a gored [weir] is a matter of strength rather than skill, for a strong wall of stones, taken from the beach and piled upon one another, is erected on the shore, until it encloses a large oval-shaped portion of the beach; the extension of the wall being usually about 200 yards or more. At the deepest point in the gored between two of the lower stones, there is an opening bridged by one very large stone supporting others. A drain is thus provided and across the drain are placed strong, slender stakes, or sometimes in these later days an iron grating.[20]

Further up the coast there were *goredi* just north of Aberystwyth (*Gored Wyddno*);[21] at Tywyn, near the mouth of the Mawddach, and at Aberdaron, while in south Wales, Swansea Bay, especially the Oystermouth-Mumbles district, had as many as thirteen stone and wattle weirs operating in the late nineteenth century.[22] These weirs were about 6 feet high and were composed of stakes driven into the sand and wattled so as to constitute a fence. In describing the Swansea Bay weirs Matheson says:

> The arms of the weir formed an angle of approximately ninety degrees and each measured up to 200 yards in length. At the junction of the arms, near low water mark, was a closely woven conical basket with its entrance facing the inner side of the weir. In order to retain some water when, during spring tides the tides ebbed beyond the limit of the weir, a layer of bushes and matting extended, close to the ground, along each arm of the weir, to a distance of about 50 yards from the basket.[23]

Sometimes a continuous trap of two or three miles in length would be formed by a line of these weirs, the inner ends of the adjacent arms being in some cases united. Complaints about the destruction of young fish by these so-called fixed engines were frequent, and in some cases nets were substituted for the wattled fence, this being said to result in the capture of fewer small fish. Of course, the Swansea Bay weirs were not specifically designed for the capture of salmon and other migratory fish, but being fixed and solid structures they were able to trap all sizes and types of fish. The remains of the weirs may still be seen, for the construction was such that they were made durable by the silting of sand and stones around them. They were in use until the end of the nineteenth century, but were gradually supplanted by stake-nets that persisted in Swansea Bay until the late 1930s.

In north Wales, especially in the Menai Straits, fishing weirs were commonplace and a weir at Penrhyn Castle, Bangor, near the mouth of the Ogwen, was in use in recent years. In this weir, there was no special cage for the fish but they were caught in the apex made by the arms of the weir. In the Menai Straits, below the city of Bangor, the remaines of *Gored y Git* may still be seen. It was referred to as early as 1588 and ended its career in 1852. It was leased in this year to Messrs Daniel and Jonathan Russell, who designed it to utilize the mud flats as oyster beds. 'They removed the stakes and formed a series of banks of stone for the protection of their oysters ... Their banks of stone ... still remain.'[24] The gored was almost certainly the one referred to by the Revd John Evans when he toured north Wales in 1798 and described it as 'a salmon snare'. It consists of posts driven into the ground at regular intervals with wattling in between. It was about 10 feet high and extended into the Straits in a semicircle about 700 yards long.

At Cored Ddu on the Anglesey shore of the Menai Straits, a weir was constructed between a small island and the mainland. Here the fish could pass round the outside of the island and towards the shore on the flood tide, but were held back by the wattled barrier of the weir as they made their way through the channel on the ebb. At Beaumaris a weir known as 'Lyme Kilne Fishery' was erected and in 1448 a second was built alongside it. The Beaumaris weirs were constructed of wattle with an arm at right angles to the main wall of the weir to provide a 'crew' or hook to prevent the egress of fish. A recess in the wall at the apex of the two arms could be closed or opened by a wooden trapdoor. The Beaumaris weir was operating in the 1920s and can still be seen near the lifeboat station, while another at Ferryman Warth, Beaumaris, has virtually disappeared. The remains of two other weirs can still be seen on the foreshore at Beaumaris.

Among other weirs in the Menai Straits were those at Borthwen Ferry House, at the mouth of the Cadnant and at Treborth. The last was located on the Caernarfon shore and was removed when the Britannia railway bridge was constructed. The most complete of all the weirs in the Straits is that of *Ynys Gomd Goch*, described in detail by David Senogles.[26] This is a double weir built from an island between the two bridges in the Menai Straits and it is capable of catching all types of fish. The fish are kept in the weir, not so much by the half-circular shape of the arms but by the strength of the tide, and after a short time they are retained behind two 8-foot ramps at the inner end.

Drifting and trawling

Undoubtedly most of the inshore fishing in Welsh waters throughout the centuries was carried out in boats of varying sizes and design, equipped with nets that were especially adapted by fishermen for capturing the particular type of fish found around the coast. Of course a number of push-nets and stake-nets as well as the widely used shore seine have been extensively employed, but since most of these instruments had to be used in close proximity to the shoreline, the amount of water that could be effectively covered by such devices is strictly limited. By using a boat, there is no such limitation and the fisherman is able to search for his catch over a greater expanse of water; he can transport his equipment from place to place according to the movement of shoals, to obtain the best possible results.

For the capture of pelagic fish, such as herrings, sprats and mackerel, the most common type of net used in the past was the drift-net. This consists of a wall of netting attached to a boat or series of buoys and free to move with wind and tide. It is a gill-net and the mesh is of such a size that the type of fish being sought will swim part way through and then be meshed by its gills. Fish that are too small will pass through the net and those too large to get their heads into the mesh will bounce off. Drift-netting is therefore a very selective method of fishing and it is usual to use nets of different mesh according to the size of fish expected and according to by-laws. Although in Wales, drift-nets were used primarily for the capture of herring as described in Chapter 3, they were also used for other fish. In the Wye, Usk and Clwyd estuaries, for example, drift-nets are still used for salmon and migratory sea trout[27] while the complex triple-walled trammel-net of the Dee estuary, used by many generations of Flint fishermen may also be regarded as a variation of the drift-net.[28] In other areas such as the Llŷn peninsula; trammel-nets (*rhwyd dramel*) were used occasionally for the capture of rays, skates and other fish and the nets were usually attached to a pair of buoys to drift with the tide.

Drift-nets can be used from small rowing-boats or from steam or motor drifters; they can either be attached to the boat or left unattended to swim free until the fisherman returns at a later time to clear them. They can be used singly or in 'fleets' of 50 to 100 nets. A typical drift-net used in Cardigan Bay, principally for the capture of herrings from motorized drifters, measures 60 yards long with corked headline and lead-weighted foot-rope and is usually between 6 and 12 score deep; a score being the customary measure of 20 meshes used to indicate the depth of a drift-net. Of course the mesh of the net can vary according to the type of fish that the fisherman wishes to catch. Thus a herring net has a very small mesh while nets for the capture of such fish as cod, skate, bass, mullet, rays, salmon and sewin have much larger meshes. A cod drift-net may have a mesh of from 7 inches to 8 inches and for rays as much as 18 inches. Until about 1950 most drift-nets were made of cotton treated with cutch, alum and linseed oil, sometimes stiffened with tar. Since that date synthetic materials such as nylon have virtually replaced the old natural fibred nets.

The boats used for drift-netting vary tremendously, and although small rowing-boats equipped with a net or two were commonplace along the whole Welsh coast, by far the most popular type of vessel was the Scottish type of drifter, 45 feet or more in length with wheelhouse, accommodation and engine room aft with a spacious deck forward of them. In the middle of the foredeck was a rectangular hatch over a hold that carried nets and the catch of fish. Generally the drifter was equipped with a mainmast forward of the hatch and always a mizzen complete with sail stepped aft. The mizzen sail was always set when fishing to keep the vessel's head up into the wind.

Today small, diesel drifters still operate on a greatly reduced scale from many Welsh ports, but especially from Conwy and Bangor. However with the decline in herring fishing, drift-netting is not as widely practised as in the past and other methods mainly for the capture of demersal fish have become far more widespread.

Although the shore seine is widely used in estuarine waters in Wales, the so-called 'tuck-net' which is used for seining in deep water was never of any consequences in Wales. For the efficient performance of a tuck-net, tides should be weak and the ground relatively clear and since no section of the Welsh coast displays those characteristics deep-water seining has never been of importance in Wales as it was in Scotland.

The most important, Widespread and effective method of catching demersal fish is the trawl-net. The trawl is

... a conical bag of net with a wide mouth that is kept open while the vessel is under way. The methods of keeping the mouths of the net open are of two kinds and these are given the names of the two types namely the 'Beam Trawl' and the 'Otter Trawl' ... In the former the mouth is kept open by a rigid spar or 'beam' of timber supported at the ends by the iron 'trawl heads'; in the latter the effect is produced by two 'otter-boards' or 'doors', solid wood and iron structures, which, being towed by warps attached rather forward of their centre, tend to diverge ... and produce and maintain the width of the mouth of the net.[29]

Both types of trawl-nets have been widely used in the Welsh fishing industry and the trawl-net is one of the most efficient instruments for the capture of bottom-feeding fish. Even if a vessel stops on a trawl, the mouth of the beam trawl-net is still kept open, but for the efficient performance of the otter trawl the fishing vessel must be on the move. Beams can be as wide as 50 feet but their use is limited to small inshore vessels concerned with catching a range of fish from shrimps to whiting. For deep-sea trawling from such ports as Milford Haven, the beam trawl has been superseded by the otter trawl. The method of trawl fishing has been described as follows:

> The net of both forms of trawl consists of a conical bag of netting which tapers to the 'cod-end', which is the main receptacle for the fish, from the forward section, which may be termed the collection portion. The otter trawl is constructed in two parts – the 'upper' and 'lower' – which laced together form the completed net. The 'upper' part consists of the upper wings, square, baitings, and upper cod-end; the 'lower' part comprises the lower wings, belly and lower cod-end. The upper wings and the square are attached to the 'head-

Hauling in the cod-end on the Cardiff trawler Iwaite, *1946 (Photo – J. K. Neale)*

rope', while the lower wings and the belly are attached to the 'foot-rope'. These two ropes enclose the mouth of the net, and their ends are attached to the 'dan lenos' or the otter-boards. Occasionally in the forward section of the cod-end there is inserted a piece of netting which acts as a non-return valve and prevents the fish from escaping: this is known as the 'flapper'.

The trawl is kept open by two otter-boards, or 'doors', which when towed along the bottom act like kites and spread the net. Restrained by the wings of the net or, in the case of the Vigneron-Dahl type of gear by the bridles running from the 'dan lena' attached to the wings of the net, they are dragged at an angle of from 20 to 30 degrees from the direction of tow. The size of the otter-boards varies according to the size of the vessel and the gear used but on an average they are from 8 feet by 4 feet to 11 feet by 5 feet in the bigger trawlers. A trawler tows at a speed of 2 to 3½ knots over the ground. For rough grounds wooden or steel bobbins or rollers are fitted along the ground rope to reduce chafe and, in order to raise the mouth of the trawl, metal or plastic floats are fitted to the head rope. Each otter-board is shackled to a steel warp and each warp leads to a separate drum on the trawl winch. When towing, the length of warp is usually three times the depth of water in which fishing is taking place. As much as a mile of warp may be carried on each drum of the winch; each is marked to ensure the gear fishing square.

There is an increasing tendency for modern vessels to be constructed for stern-trawling, one main advantage being that the open after deck is protected in bad weather by the bridge structure as the trawler is generally head to wind when shooting and hauling. The net is practically the same as that for the conventional method except that sometimes it is fitted with two cod-ends.

To shoot the net the vessel is headed into the wind and speed maintained at about 2 knots and the net, which is ranged along the open after deck, is attached at the cod-end to an outhaul wire, which hauls the net to the stern ramp. The outhaul is disconnected and connected to the net further forward at about the position of the foot-rope and the net hauled further outboard. The cod-end is now trailing in the water sufficiently to drag the net completely into the water under the control of the warp winch. When the dan lena reaches the end of the ramp the winch brakes are applied and the back strap legs of the trawl boards clipped to the cables. Outhauls are disconnected, speed temporarily increased to 6 or 7 knots and the trawl is shot away. Whilst paying out the last 50 fathoms the speed is reduced to trawling speed -about 3 knots.

On completion of the tow the gear is rapidly have in and virtually the reverse process outlined above is gone through until the foot-rope is aboard. The gilson is then attached to a becket on the cod-end which is have aboard and the catch emptied into the fish pounds on the deck below through a hydraulically-operated deck hatch.

The conventional side-trawler differs greatly in the method of shooting to that of the stern-trawler. Here the trawl is shot to the windward side so that when it is in the water the vessel will drift away from it, thus streaming the net clear of the ship's side. Both doors are lowered to the water's edge and the aft and fore warp are paid away a few fathoms, the fore warp being given an addition amount of length to compensate for the distance between the fore and aft gallows -iron frames fitted with a sheave and pulley through which the warp is streamed. The vessel is then set full speed ahead, the helm being set to run the vessel in a circle, or segment of a circle, with the centre towards the gear. When the required length of warp is paid out, the 'Messenger', a heavy iron hook, is passed round the warps; the inboard end is then passed round the winch drum and hauled in, thus pulling the warps together into the towing-block. Minor adjustments are then made to make sure that the warps are even and the engines are rung down to towing speed.

After towing from twenty minutes to three hours, according to the size of vessel or the state of the fishing, the trawl is hauled. About ten minutes before the completion of the tow the vessel is brought before the wind. The mate takes station at the winch and the third hand at the towing-block, armed with a crowbar; on the order 'knock-out' the third hand knocks the towing-block open and the warps immediately separate, the fore warp breaking clear of the ship's side. A few turns are taken on the fore drum of the winch, the wheel is put over towards the gear and, after a short pause, both warps are hauled together, the engines then being rung off. By the time the after door reaches the gallows the vessel will usually be athwart the wind, with the gear on the windward side. The doors are brought up to the gallows sheaves, the weight of them taken by a 'stopper' and the warps eased. The quarter ropes are then unbent from the doors, passed through sheaves on the engine room casing and brought to the winch. The winch is started and the net is hauled inboard until the foot-rope is over the gunwale when the remainder is hauled by hand.

The cod-end is hoisted inboard by means of a becket and the 'gilson' which is a tackle rope from the masthead. The cod-line is unbent and the fish fall into the deck pounds.

Though possible at anything up to 300 fathoms, little trawling is carried on in depths exceeding 250 fathoms.[30]

The beam trawl was undoubtedly designed for use in sailing vessels dependent solely on wind, tidal streams and currents for their motive power. When there was no wind, a sailing vessel could make very little headway through the water but nevertheless, the mouth of the beam trawl would remain open and it would continue to take fish. It was the widespread adoption of the steam trawler and later the internal combustion engine~ that led to the gradual disappearance of the beam trawl except on the smaller vessels. It was gradually superseded by the otter trawl which had a far greater catching capacity as long as it was kept moving in the water continuously. In Wales, it was really after the 1914-18 war that the otter trawls gradually replaced beam trawls except in such places as Milford Haven, Swansea and Cardiff, for steam engines were too large and weighty to be installed in small craft and so it was not until the compact marine internal combustion engine became available that otter instead of beam trawls were used to any extent. The otter trawl had many advantages over the beam trawl:

> Its components can be handled more easily. It can be towed with wire warps which stow in a minimum of space on the barrels of a winch. Its headline is raised considerably higher off the seabed by means of floats and it will therefore catch fish further off the bottom. And if bridles are used the otter-

Trawler Skipper – Walter Rymer of the Iwaite, *1946*
(Photo – J.K. Neale)

boards will guide more fish towards its mouth than will the heads into the mouth of a beam trawl – it can be hauled aboard over the stern of the boat, either up a ramp or otherwise, as well as over the side, whereas a beam trawl has to be brought inboard over the side.[31]

Lining

Along many parts of the Welsh coastline, hand lining or trolling is carried out by fishermen principally for the capture of mackerel and a certain amount of cod. Any type of boat from the smallest can be used for trolling and a number of baited hooks are drawn through the water from the moving boat. A variety of baits ranging from mussels to lugworms and from sand eels to crabs is used and the trailing line is watched continuously by the fishermen and hauled in whenever a fish is felt. Feathers dyed in bright colours bound to the back of hooks may be good enough to attract mackerel and usually in the small boats engaged in mackerel fishing from Cardigan Bay villages, a hemp line trailing from the stern of a slowly moving boat will carry no more than fifteen hooks.

Hand lines with baited hooks may also be set at low water on the sands and fished on the next ebb. This method of fishing was once common on the north Wales coast around Conwy where the lines were known as 'tee lines'. Horsehair snoods carrying the hooks were tied to the hemp line at intervals of between 8 inches and 15 inches, and anything up to 120 hooks were used for the capture of a variety of fish; the lines being kept in place with pieces of wood each shaped into a solid figure of eight, known as 'chubb'. The lines were adaptable enough to be used for lining from a boat.

Off the Icelandic, Norwegian and Greenland coasts long lines, perhaps up to ten miles in length, are used over rough ground to capture halibut, skate, ling, cod and other large fish. In the Irish Sea smaller lines are widely used by trawlermen and driftmen for the capture of fish. These lines carrying perhaps a hundred hooks for each 75 fathoms of line had to be baited ashore, often by the women, coiled into a basket or wooden tray and placed on board a vessel. The line carrying a buoy is shot from the vessel and more buoys are anchored at intervals along the line at the extreme ebb. The line remains in the fishing position before it is hauled in and the catch removed. In the 1920s long lines worked out of Milford in search of conger, cod, skate and halibut. Each vessel carried fourteen lines, each of 600 fathoms with a 100 hooks on each line.

8 THE FISHING PORTS

Throughout the centuries, the majority of Welsh fishermen with some notable exceptions, have concerned themselves with inshore fishing more often than not within sight of land. In the past most of the activity was carried out from small boats and many a fisherman could sail away to the fishing grounds in the morning and return to a home port in the evening. Many Welsh fishermen, too, were by tradition part-timers and on many sections of the Welsh coast, sea fishing could be combined with agricultural labouring, rabbit catching and other shore-based occupations. In many coastal communities such as southern Ceredigion, herring boats were largely manned by merchant seamen home on leave from deep-sea sailing or by retired seamen. In some small communities that supported some full-time fishermen, those men by concentrating on different types of fish at different periods of the year could obtain a livelihood. Thus, for example, in the ancient borough of Conwy, until recent times some fishing families were engaged in mussel gathering in the estuary of the river in the autumn and winter months. For a couple of weeks in the spring, armed with a seine-net, the fishermen made their way upstream to Tal-y-cafn to capture the elusive sparling.[1] The short sparling season was followed by a longer period fishing for salmon and sea trout with a seine-net, while at the height of the tourist season many a Conwy fisherman hired his boat for fishing and pleasure trips from the quayside.

Of course in Wales, there were some notable exceptions to the haphazard and limited scale of fishing operations. Coastal towns such as Milford Haven, Cardiff and Swansea developed into fishing ports of considerable importance.

Cardiff

The history of trawling in Cardiff is really tied in to the fortunes of one company – Neale & West – who formed a partnership as fish-merchants in 1885.[2] Joshua John Neale, born in Ireland of English parents, and Henry West were cycling friends in Bristol. In 1885 they moved to Cardiff to establish a fish-merchants' business at Custom House Street. The fish they sold in their business came from other ports but they persuaded some of the Cardiff tug-boat masters to take beam trawls on board their vessels while seeking tows in the Bristol Channel. The fish caught by the tugmen were then sold to Neale & West. The supplies were irregular and the partners decided to purchase a second-hand trawler of 52 net tons from a Hull trawling company. The *Lark*, built in 1886 was purchased in 1888 and she was a vessel 90 feet long, 20 feet broad and had a draught of 10 feet 6 inches. She had a bridge behind the funnel but had no chart room or wheelhouse. The vessel was equipped with a beam trawl.

The company flourished tremendously and by 1906 its trawler fleet consisted of nineteen vessels based in the Bute West Dock at Cardiff.[3] In the early years, Cardiff had no ice factory, so that cargoes of Norwegian ice had to be imported until the Cardiff Ice and Cold Storage Co. Ltd. was formed by Neale & West. Two of J.J. Neale's sons, Douglas and Howard, went to work in ice-making plants in Cardiff and also at the Milford Haven Ice Company. Apparently at first there was a strong prejudice amongst the trawler skippers of Cardiff against the use of artificial ice and 'Mr Neale had actually to give the ice in some instances and pay the skippers to use it'.[4] J.J. Neale's seven sons – Wilfred, Morley, Symonds, Stanley, Nelson, Howard and Douglas – were all concerned with the family firm, either operating the trawler fleet or working in the ice-making division. In 1907 a trawler fleet WflS operated from Milford Haven as well as Cardiff and this was managed for the company by a James Thomas. Henry West was reluctant to expand the business as the Neale brothers desired and in 1910 he left the company although he continued his association with the Ice Company. Two or three of the brothers who were concerned with the Milford Haven operation brought their trawlers back to Cardiff and a new company, Neale and West Ltd., was founded. Most of the older trawlers were sold and new, larger vessels designed to pursue the hake fishery all the year round in deep water, were purchased. These trawlers – *Saxon*, *The Norman* and *The Roman* – operated by the Neale Brothers were brought back from Milford Haven. and new vessels, many of them bearing Japanese names were built especially for the company. 'J.J. Neale had become very friendly with the Japanese, possibly because of a visit to Cardiff of Japanese warships' said Mr Jack Neale:

A Mr Kunishi came to Cardiff to study Neale & Wests' methods and was given every assistance, going through every department, going on fishing trips etc. I think he was financed by his partner Mr Tamura. In due course grandfather

A Neale & West steam trawler entering Bute West Dock, Cardiff c. 1950

had a trawler built and sent her to Japan with a good crew who then trained Japanese crews to trawl as there were no trawlers in Japan previous to this ... In no time at all the Japanese firm surpassed Neale and West.[5]

In the hey-day of the fishing in 1913, Neale and West possessed eighteen vessels, all more than 200 net tons in size.[6] As many as ten of those trawlers were sunk during the First World War, when the whole fleet was recruited for mine-sweeping duties. At the end of hostilities, Neale and West were left with a fleet consisting of eight pre-war vessels and seven built between 1914 and 1920.[7] All these vessels varied from 230 gross tons to 290 gross tons, 130 feet to 140 feet long and capable of fishing in depths to 150 fathoms.

Up until 1920, the type of trawl used by the Cardiff trawlers was the otter trawl, but in 1920,

... the Cardiff firm – the first in the country to do so – adopted a type of trawl known as the Vigneron-Dahl, or more commonly as the 'V.D.' or 'French trawl'. One of their skippers was sent over to France to try the new gear, and as a result of comparative experiments with this and the ordinary trawl, the Vigneron-Dahl gear was adopted and is now used on all the Cardiff trawlers. Most of the other western ports also use it today, though it is not so common in the North Sea. The main distinguishing features of the 'French' trawl are as follows. Instead of the otter-boards being attached directly to the end of the wings which are longer than in the ordinary trawl, they are separated from the latter by long 'ground warps' of wire and manilla, from 30 to 60 fathoms long (60 to 90 in the case of the Cardiff boats). The ends of the wings are attached to steel or wooden poles, and the head-line is buoyed by a line of glass balls, set most closely together towards the centre. As a result of these modifications the apparatus would appear to work a wider area than the ordinary trawl, since the otter-boards are farther apart when towing, and the ground warps, being close to the bottom, would tend to scare the fish inwards towards the oncoming net. It is claimed that the use of the 'V.D.' trawl gives fish in better condition and in greater quantities and that it reduces the trawler's coal consumption.[8]

In pre-1914 days Cardiff trawlers went as far as the coast of Morocco in search of hake but in the 1920s the fishing grounds off the west and south coasts of Ireland were those fished by Cardiff trawlers. Matheson provided details of a typical voyage of the SS *Ijuin* in 1923. This vessel with its crew of master, mate, wireless operator, two engineers, two firemen, cook and four general hands spent about two weeks on a fishing expedition:

Left Cardiff 11 a.m. on Tuesday, September 4th. Steamed to about 5 miles off north end of Lundy Island, from there setting course for Fastnet Rock, S. Ireland, West and North West, distance 190 miles. Passed Fastnet about 1.30 p.m. on Wednesday, September 5th, and altered course to N.W. for Bull Rock, distance 30 miles. Arrived off Bull about 4.30 p.m. and altered course to N.N.W. Steamed on that course for 105 miles. Stopped the ship about 4 o'clock a.m. September 6th and took soundings, result being 180 fathoms. Shot the trawl, and for three days fished between latitudes 52°25′ N. and 52° 50′ N., in different depths ranging from 100 fathoms to 220 fathoms. Very medium class of hake, highest catch being eighteen score and lowest five score. On the night of September 8th steamed 95 miles N.N.W. to the N.W. edge of the Porcupine Bank. Fished there for forty-eight hours, best catch being ten score of medium hake. Considerable number of flatfish and bream generally caught here. Greatest depth fished in, 160 fathoms. September 10th, steamed 110 miles E., arriving off Achill Head and Black Rock on N.W. coast of Ireland. Fished here between latitudes 54° 10′ N. and 53° 55′ N. in 110 fathoms until the night of September 12th. Plenty of fish here, some ships catching as much as thirty score of medium hake and twenty baskets of small hake. Experienced a considerable amount of 'rumpy' [torn nets], losing two complete trawls, and were compelled to leave. Steamed 60 miles S.W. and fished all day on September 13th in latitude 53° 10´, depth 100 fathoms. Different class of hake altogether, very large but very soft. Steamed again on the night of the 13th about S. by E. sixty miles. Had one haul on September 14th about twelve miles off the Great Skelligs Rock, sixteen score very large hake. Left for home and arrived Cardiff seven a.m. on September 16th.

In the 1920s and '30s, the Neale & West fleet was completely modernized with vessels of about 300 gross tons being purchased from shipyards in Selby, Middlesbrough and elsewhere.[9] The *Kunishi* of 1927, for example, was built at Smith's Dock, Middlesbrough, and was a vessel of 303 gross tons, 130 feet long, 24 feet wide and had a draught of 12 feet. The *Akita* of 1939 was the same size and was built for Neale & West by Cochrane & Sons Ltd. of Selby, Yorkshire.

The Cardiff trawling industry, although it recovered slightly after the end of the Second World War with the replenishment of fish during the war years when relatively little trawling was carried out, soon fell on hard times. In 1950, Neale and West still operated thirteen vessels, all except one being of pre-war vintage.[10] That post-war vessel was the 361 ton St Botolph built in 1946 by Cook, Wellington and Gemmel Ltd., Beverley, Yorkshire. By the early '50s the hake grounds had become depleted and Neale & West were experiencing great difficulty in recruiting suitable staff. Many of the vessels had been sunk during the war, but their replacement was second-hand tonnage, built in the '20s and used for many years by other trawlermen. By 1955 the company was in deep trouble; fish stocks were greatly depleted; the cost of bunkers and other necessary supplies was prohibitive and there was also 'the major problem of crew misbehaviour'.[11] The trawlers themselves had to sail further and further from their home port in search of fish and those trawlers were not large enough or modern enough to sail to distant Arctic waters to fish. In 1956 the *Sasebo*, *Akita*, *Chaffcomb*, *Oku*, and *St Botolph* were sold to the Boston Deep Sea Fishing and Ice Co. Ltd. The *Nodzu* was sold to the Dalby Steam Fishing Company of Fleetwood and the remainder went to the shipbreakers' yard. The year 1956 therefore saw the end of Cardiff as a fishing port of any consequence and trawlers no longer came to the quayside in the West Bute Dock.

Swansea

Although Swansea Bay was a noted centre of oyster fishing and set-netting, it was not until the end of the nineteenth century with the widespread adoption of steam for the propulsion of ships that it attained some importance as a deep-sea fishing port. Fish was certainly landed at the port in the late eighteenth and early nineteenth centuries and a fish market was opened in 1792.[12] Just as the fortunes of Cardiff as a fishing port were monopolized by the firm of Neale and West, so too was the development of Swansea as a fishing port in the hands of the Castle Stearn Trawling Company. This company was established at Milford Haven during the last decade of the nineteenth century but in 1903 they moved their centre of operations to the South Dock in Swansea. In that year the 15 trawlers, all of less than 60 net tons owned by the founder G.H.D. Birt of Milford Haven, passed into the hands of Crawford Heron who almost immediately transferred the fleet to Swansea.[13] During the first decade of the twentieth century other trawler owners such as Alex Keay of the Hector Stearn Trawler Company with a single vessel, the *Carlam*, and A.J. Newman who purchased the Cardiff trawler *Labore et Honore*, operated from Swansea, but it was the Castle Steam Trawlers Ltd. that dominated the industry. The old Milford trawlers were scrapped and Crawford Heron invested very heavily in new vessels averaging 250 gross tons. No fewer than 10 new vessels were built between 1905 and 1908 and at the outbreak of the First World War, the Castle fleet consisted of 19 steam trawlers.[14] Those vessels were not only concerned with fishing in the Bristol Channel and off the Irish Coast but they also visited distant grounds such as the Burlings abreast of Portugal.[15] The progress of Swansea before 1914 was rapid. Among its geographical advantages were proximity to the coalfields which provided the large tonnage required for bunkering the steam trawlers and its railway facilities for despatching large quantities of fish daily to the thickly populated adjacent valleys. New vessels were regularly built and added to the Swansea fleet and in the evolution of the steam trawler the 'Castle' type of trawler became well known in the fishing industry as a vessel suited in all respects for the particularly exposed conditions of the westward fishing grounds.[16]

As with the Cardiff trawler fleet, the Swansea vessels were taken over by the government in 1914 for minesweeping duties and six were lost. At the end of the war Crawford Heron's fleet was taken over by one of the largest trawling companies in the United Kingdom – the Consolidated Steam Fishing and Ice Co. Ltd. of Grimsby which owned as many as 124 trawlers. Curing houses were erected at the South Dock in Swansea and the fleet of trawlers grew rapidly: in 1929 'practically the whole of the South Dock Basin is given over ... to the coaling, icing, storing, and equipping of the stearn trawlers.'[17] The twenty-five or so vessels operating from Swansea were Grimsby registered and the headquarters of the company, known as Sir John D. Marsden Bart., was Auckland Road in the Grimsby Fish Docks. In the 1920s and '30s Marsden's Swansea trawlers were forced to go further afield in search of fish and the new vessels the company built were large enough to venture to the more exposed part of the Atlantic. They fished off the west coast of Ireland, the Northern Channel off Scotland and the English Channel and also went as far as the coasts of Morocco, Spain and Portugal. The Swansea fish market was of course of considerable importance and in addition to handling the catches of its own fleet, visiting trawlers used the port and its facilities as their base during the fishing season. In 1924 no fewer than 626 men were fully employed on Swansea trawlers and an average of thirty-two vessels operated from the port between 1925 and 1939.

After 1945 the Swansea trawling industry never really recovered and Consolidated Fisheries Ltd. diverted some of its trawlers to Milford Haven. The *Picton Castle*, a vessel of 309 gross tons worked from Milford until the 1970s and although today a certain amount of inshore fishing is carried out from Swansea, the trawling industry that brought so much prosperity to the port has ceased. In 1979 only 8183 tonnes of fish, principally skate, was landed: in 1980 – 102.23 tonnes: in 1981 – 152.58 tonnes: in 1982 – 167.56 tonnes and in 1983 – 172.89 tonnes.

Tenby

One of the earliest and most noteworthy fishing ports in south Wales was Tenby and the Welsh name of 'Dinbych-y-Pysgod' (Tenby of the fish) points to the importance of fish in the economy of the port. Tenby's pier is said to be the earliest in Wales and 'It was in 1328 that Edward III gave Tenby the right to levy dues to pay for its building ... it protected the anchorage from the north easterlies.'[18] This curved pier was replaced in 1842 by the present straight pier and in the same year St Julian's Chapel that stood on the pierhead was also demolished, but a new chapel was built at the head of the harbour in 1878. In the eighteenth century 'prayers were said there for the fishermen and their boats. A tax of a halfpenny was paid to the officiating priest by each fisherman and a penny for each boat'.[19] Obviously

Tenby was a major fishing port and as a centre of inshore fishing it pre-dated Milford Haven as the premier fish port of south-west Wales by many centuries. Fish ranging from herrings to oysters were landed at the port, and the seas between Worms Head in Gower and Caldy Island were particularly productive. The fishing ground most frequented by Tenby fishermen before the seventeenth century was known as 'Will's March' a bank that lies about ten miles south by east from Caldy Island.[20] Until the end of the nineteenth century, Tenby fishing boats limited their activities to Carmarthen Bay, to an area bounded by Worms Head, St Govan's Head and Lundy Island. During the twentieth century some Tenby boats went further afield and searched off the south and east coast of Ireland and the approaches to the Bristol Channel: 'Those trawlers were found off the Wexford coast from November to Christmas; between Lundy and the Longships from January to April. They returned to refit at Easter and during the summer worked their own fishing grounds in, or just outside Carmarthen Bay.'[21] Although there was a great deal of fishing activity in Tenby over the centuries and inshore fishing and oyster dredging was practised on a considerable scale by the mariners of the town, the industry never developed on the scale that it should have done, due to the poverty of the fishermen. Thus a great deal of the fishing carried out from the port was in the hands of Brixham fishermen who for centuries visited Tenby annually in their trawlers during the summer

Tenby Luggers. Susan Ann and Mumbles in harbour c. 1900

months. Some of those Brixham men actually settled in Tenby. Gradually during the nineteenth century, locally-owned trawlers and drifters increased considerably and with the arrival of the railway in 1866, it became possible to send fish to the industrial valleys of south Wales. By 1870 Tenby with its own fleet of trawlers, drifters and oyster skiffs and the visiting Brixham trawlers was the only important fishing port in south-west Wales. Beam trawling for a variety of fish was carried out by a dozen or so trawlers that operated from the port. By 1891 a total of nineteen vessels up to 50 tons in size was based at Tenby. Of these vessels, sixteen were cutter rigged and three ketch rigged. Some of the trawler masters owned their own vessels others were paid on a share basis with the rest of the crew. The large trawlers carried a crew of three men and a boy, or in a few cases two men.

For line fishing for cod between November and February, open boats were used. In 1864 as many as thirty of these deep-keeled, two-masted luggers were employed during the season. By 1891 as many as forty-nine luggers between 15 feet and 25 feet in length were employed in line fishing, together with a certain amount of oyster dredging and net fishing.

By the last decade of the nineteenth century, the fishing industry of Tenby was on the decline and the ancient port was developing rapidly as one of the main holiday resorts of Wales. Pleasure boats replaced the traditional fishing craft and Tenby could no longer compete with the rapidly expanding steam trawler fleet of nearby Milford Haven. In 1896 only 32.59 tons of fish were landed at Tenby and by 1914 this had declined to a mere 7.9 tons. Today Tenby does not figure in the official statistics of fish landings and the industry that once provided the town with one of its main activities has virtually ceased to exist.

Milford Haven

Milford Haven became a fishing port almost by accident. The maritime connections of the port located on the north shore of a great natural inlet, go back for many centuries, but it was really the last two decades of the nineteenth century, following the opening of the docks, that ushered in a period of great prosperity. In previous years there had been many ambitious plans to develop Milford as a large naval dockyard and as a major trans-Atlantic terminal that was to outstrip both Liverpool and Southampton as the great port for oceanic liners. In the early 1900s, the harbour was the home of a group of refugee whalers from Nantucket, but their use of Milford as a whaling port was short-lived.

Local entrepreneurs during the last quarter of the nineteenth century saw the trans-Atlantic passenger trade as the only future for Milford but after many setbacks to their ambitions the Dock Company reluctantly came to recognize that the future lay in the development of the fishing industry. 'The Dock Company have not built their great docks to attract fishing smacks', said a *Times* correspondent in October 1899, but 'they will not look askance on the great possible development of the fish trade: In his speech to the half-yearly meeting of the Milford Docks Company in July 1890, Thomas Wood, the Chairman, pointed out: 'That much derided and despised fish trade has come in very opportunely for us and yields us a very considerable amount of revenue ... It is a trade we did not either cater for, or look forward to ... but it helps to pay, and in fact does pay the expenses of the docks.'[22] What had been thought of in derisive terms as an industry that was tolerated on sufferance, was soon developing into the staple industry of the port. Although the large passenger liner, the *City of Rome*, anchored in the Haven, six miles from the Docks in 1889, it did not herald a great period of trans-Atlantic traffic. Gradually people were beginning to realize that Milford would have to rely on the despised fishing trade 'and give up its dreams of becoming an ocean port and be content with the more commonplace though no less useful role of a Welsh Grimsby'. Nevertheless, abortive attempts at greatness still persisted. In 1889, for example, it was announced that a fortnightly service, becoming weekly, was to be set up between Canada and Milford. The inaugural ship the *Gaspesia* reached Milford in December 1898; her complement of seventy-five passengers (her capacity was over 600) boarded a special record-breaking train for London and the Docks Company congratulated themselves on their great victory. Neither the *Gaspesia* nor any other trans-Atlantic liner visited Milford ever again and for the next fifty years or so it was the fishing industry that was to dominate the economy and life of Milford Haven.

The first vessel to enter the new Milford Docks in 1888 was the steam trawler *Sybil* and this ushered in a period of great prosperity with many trawlers joining the Milford fleet every year. In south-west England Brixham declined as a fishing port and the once rich fishing grounds off the Cornish and Devon coasts were thinning out. Fishing vessels from Brixham and from other ports were attracted to Milford to take advantage of the major fishing grounds of the Western Approaches. During the last decade of the nineteenth century the quantity of fish landed at the port more than trebled, being 5,610 tons in 1890 and 18,245 tons in 1899. By 1889, twelve trawlers with a total tonnage of 1,100 tons were using the port and thereafter the development was very rapid indeed. By 1904, 28,000 tons of fish were landed at Milford and the fishing fleet in 1904 consisted of sixty-six trawlers and 150 smacks. By 1908 the total landings amounted to 44,289 tons and by that time the Milford trawler fleet was wandering further afield as far as the west coast of Spain and Morocco. Sailing trawlers persisted in the Milford fleet until after the First World War and many a conservative fleet owner believed that steam could never replace sail on trawling vessels. Nevertheless, the first decade of the twentieth century was really the age of steam in Milford. The introduction of steam trawlers increased the power and range of the fleet and the use of ice from one of the three Milford ice-making factories ensured that the catch arrived in port in good condition. Trawlers could now spend considerable periods in the fishing grounds and the widespread adoption of the otter trawl to replace the beam trawl contributed to even greater efficiency.

But all was not well in the Milford fishing industry in the years before the First World War; increased efficiency contributed to overfishing and there was a marked depletion in the fish stocks of St George's Channel. New varieties of fish that had hitherto been of little interest to fishermen were now landed at the Milford fish dock rather than being thrown back in the sea; under-sized fish, particularly hake, that were once regarded as too small to be saleable were landed in increasing quantities. Between 1900 and 1914 the total landings of fish fell very gradually although in 1911, Milford was still regarded as one of the most important fishing ports in England and Wales.

Selling fish in the Milford Haven fish market in the 1920s. The old open-sided fish market was demolished in 1991

During the first decade of the twentieth century, an attempt was made to attract to Milford drifters concerned with inshore fishing for mackerel and herrings. A new 'Herring Wharf' outside the main dock was constructed and fish could be landed here at almost any state of the tide without having to open the lock gates and cause delay in the arrival and departure of the drifter fleet. A herring and mackerel market was built alongside the quay and a small house for kippering was also built. In February 1905, seventy drifters regularly used the Herring Wharf.[23]

The extensive fish market with its complex of railway lines was completed in 1908 and this huge building, 950 feet in length, was a hive of activity. One of the essentials of a thriving fishing port is a plentiful supply of ice and in its early days the Milford fishermen depended on natural Norwegian ice from a hulk in the docks. This soon became inadequate and in 1896, Messrs Neale & West, the Cardiff trawler owners and ice makers, opened an ice-making factory in Milford. Four years later one of the foremost trawler owners in Milford, Messrs Sellick, Morley and Price, opened a second ice factory on the quayside while yet another was opened at Neyland by the Neyland Ice Company. The two Milford ice factories could not cope with the demand and Norwegian ice was still being imported to be used at peak periods, such as the mackerel season. The Docks Company were anxious to build yet another ice factory to cope with peak demand but the two existing companies objected violently and in 1912, Sellick, Morley and Price who not only owned an ice factory but also thirty-six of the seventy trawlers of the Milford fleet threatened to take all their trawlers to Fleetwood. Seven of the trawlers did leave for Fleetwood in September 1912 but the complete removal of the fleet was averted. 'The action was said to be vindictive ... The move was said to be purely dictated by economic interest. Fleetwood had made great strides as a fishing port, because it had easy access to a large population; costs there were lower and prices were said to be better.'[24] Fleetwood in 1912 had three times as many boats as Milford although the early history of both ports was somewhat similar. Fleetwood like Milford had always seemed to be on the point of prosperity but never quite made it until the fishing industry was developed. The migration of Milford's most important trawling fleet in its entirety was averted and the Cardiff Pure Ice and Cold Storage Company – Neale & West's ice factory – was extended.

Another crisis in the port during the first decade of the twentieth century was the removal of the whole 'Castle' fleet of steam trawlers to Swansea in the summer of 1903. Their departure was soon made good by the arrival of new ships and new owners. An attempt was also made by neighbouring Neyland to revive its fortunes as a fishing port with the removal of the Irish packet service to Fishguard in 1906. The directors of the Milford Dock Company seized the opportunity of this possible rivalry on its very doorstep by spending money in Milford itself to build a slipway and improve the fish market.

On the eve of the First World War, the following trawler owners were operating from Milford:

G.H.D. Birt (Phoenix Trawling Company)	*Hibernia* (built 1907)
	Louise (built 1907)
	Marion (built 1906)
	Raywernol (built 1912)
	Fishergate (built 1905)
	Gillygate (built 1905)
G.H.D. Birt & D.J. Davies	*Kirkland* (built 1908)
Edward Brand	*Cygnet* (built 1893)
	Halcyon (built 1893)
	Koorah (built 1912)
	Osprey (built 1893)
	Teal (built 1893)
Davies & Grand	*Dartmouth* (built 1896)
John H. Dove	*G.M.* (built 1910)
W. & G.P. Francis	*Mary* (built 1907)
M.E. Grand	*Cuckoo* (built 1894)
Hancock & Harries	*Dinas* (built 1909)
	Solva (built 1909)
Hancock, Harries & Coope	*Tenby* (built 1913)
Thomas G. Hancock	*Marlus* (built 1911)
Morgan W. Howell	*Gwenllian* (built 1911)

James & Longthorpe

Gloria (built 1907)

William Jenkins

Dewsland (built 1907)

Oliver Johnston

Cameo (built 1907)

David G. Jones

Alert (built 1896)
Emerald (built 1905)
Ruby (built 1907)

John Jones

Cambria (built 1905)
Dowlais (built 1896)

C.C. Morley (Southern Steam Trawling Co. Ltd.
of Milford Haven and Waterford) originally Sellick,
Morley & Price

Kalmia (built 1898)
Lobelia (built 1898)
Magnolia (built 1898)
Othonna (built 1899)
Petunia (built 1899)
Rose (built 1904)
Syringa (built 1905)
Tacsonia (built 1905)
Uhdea (built 1906)
Weigelia (built 1911)
Xylopia (built 1911)
Yucca (built 1912)

Morley & Price

Essex (built 1906)
Macaw (built 1909)

Neyland Steam Trawling Co.

Angle (built 1908)
Apley (built 1908)
Bush (built 1908)
Caldy (built 1908)
Hero (built 1907)
Slebech (built 1908)

Sydney Morgan Price (Western Steam Trawling Co.) *Abelard* (built 1909)
Alnmouth (built 1912)
Avonmouth (built 1890)
Charmouth (built 1910)
Exmouth (built 1912)
Falmouth (built 1909)
Lynmouth (built 1892)
Portsmouth (built 1903)
Sidmouth (built 1906)
Weymouth (built 1902)
Yarmouth (built 1907)

James Thomas *Beatrice* (built 1906)
Edward VII (built 1907)
Victoria (built 1912)

During the First World War, a large proportion of the Milford fishing fleet was commissioned for minesweeping and other duties, although their place was taken by trawlers from East Coast and Belgian ports. Trawlers from the Belgian port of Ostend were particularly important. With the revival

The Milford fishing fleet entering the docks in the 1920s

of fishing after the fallow years of the war, Milford flourished once again and over 46,000 tons of fish were landed in 1920. All seemed well and prosperity reigned once again and trawler owners invested in new vessels which, with hindsight, were purchased at greatly inflated prices. In 1925, Milford had a fleet of 110 steam trawlers in addition to the large number of drifters that used the port. This was a period of optimism; a second smokehouse to deal with the increased catch of herring was built in 1923 and another in 1924. These did not prove sufficient and two more smokehouses were built in 1925. Milford ranked fourth in importance after Hull, Grimsby and Fleetwood. As in Cardiff, the Vigneron-Dahl fishing equipment was introduced into Milford in 1923 and proved very efficient, although contributing to the overfishing of St George's Channel. Writing in 1929 Matheson[25] points out that 'there are always about twelve hundred men afloat in vessels owned by Milford firms. Most of the fish brought into Milford is caught at distances up to four hundred miles out in the Atlantic ... The supply of coal necessary for these vessels is considerable as each deep-sea Milford trawler burns on average about eight tons of coal per day'.

The Ice Factory and Mackerel Quay before demolition and the re-construction of the Milford Haven dock in 1991

In 1924, Milford had as many as 101 steam trawlers and eight sailing trawlers and employed over 1,000 fishermen.

During the second half of the decade 1920-30 all was not well in Milford. Trade had bately recovered from the 1926 strike and many of the owners who had purchased trawlers at the greatly inflated prices of the post-war boom were in difficulty. Several trawling companies closed down and one company, the large Iago Steam Trawling Company with its fleet of ten large trawlers, moved to Fleetwood. The fishermen's strike of 1932 did not help matters, although in that year Milford still possessed a fleet of 108 trawlers with 150 others visiting the port for eight or nine months of the year. In June 1931 another cloud, in the form of an armada of Spanish trawlers, appeared and since many of those landed their catches at Milford at much lower prices, this affected the Milford trawler owners. In 1934/5 two large trawler owners – David Pettit[26] and Brand & Curzon[27] – were forced to retire from the unequal struggle and thirty-three fewer vessels, all of post-1918 date operated from Milford. By 1939 Milford Haven was in considerable trouble due to foreign competition and the results of over fishing. Again the fishing grounds recovered during the Second World War, with the result that as much as 59,000 tons of fish was landed in 1946: an all-time record.

Since those halcyon days, the fishing industry in Milford has declined alarmingly although in June 1953 the fleet of steam trawlers still consisted of seventy-eight vessels. Grounds were overfished in the 1950s and foreign trawlers, especially from Spain, contributed greatly to the great reduction in fish stocks. The hake, for which Milford was well-known, proved to be an elusive fish, although in the early 1960s, the largest trawling company obtained some trawlers that were specifically designed to pursue the hake. 'It was found necessary [in 1964] to transfer four of these vessels to other ports. During the latter half of the year, the large vessels remaining at the port worked the haddock grounds off the south west.'[28] By 1974 the trawling fleet had declined to a mere fourteen vessels; six of which were owned by Norrard Trawlers Ltd.[29] Most of the others were owned by small companies in possession of one or two trawlers. G. Antoniazzi was typical of these small owners, operating a 1928-built trawler of 108 tons – the *Lord Rodney*. Hubert Jones owned three – the *Brenda Wilson*, *Georgia Wilson* and the *Sally McCabe*; the last operating mainly from Swansea. Since the mid-1990s, despite the construction of a new market, only about half a dozen small drifters and trawlers are Welsh owned. The major part of all landings is from French, Spanish and Belgian vessels and much of the catch is taken by road in large refridgerated lorries to the European market.

Certificate of Competence as a skipper of a fishing vessel

Excluding mackerel, caught mainly by freezer trawlers from other ports and dispatched to West Africa from Milford, the industry by 1979 only landed 1,295 tonnes. By 1983 that figure had declined to 1,062 tonnes; skates and rays being by far the most important fish landed. Nevertheless that low figure still represented 78 per cent of all whitefish landed in the whole of Wales and the declining port of Milford still contributed a lion's share to an industry that itself has witnessed a general decline in all parts of the United Kingdom.

In 1985, six part-Spanish part-British fishing boats landed fish at the Fish Quay. About four local trawling companies operated from Milford Haven, two being part of a Fishing Co-operative in the port. The picture was ever-changing as old vessels were scrapped and new tonnage brought in.

In 1991, the future of Milford Haven was seen as lying in tourism rather than the fish trade. Acting as host to the 'Tall Ships Race' in July 1991, the dock was converted to a pleasure marina. The extensive fish market was demolished, the Old Mackerel Quay was converted to a landing place for pleasure craft; smokehouses and ice factory disappeared and the half dozen or so steam trawlers ceased to operate, probably permanently, with the bankruptcy of the local trawler company. Nevertheless a smaller ice-making plant and a new fish market was constructed on the West quayside, but whether those modern facilities will ever be used again by a Milford Haven trawler fleet is doubtful. A few small locally owned inshore fishing boats are still operational, but most of the fish landed at Milford in 1991 is from Spanish-manned diesel vessels that belong to other ports.

The other harbours in Milford Haven, Neyland for example, never developed to the extent that Milford itself did as a fishing port. Neyland had its ice factory and fishing fleet for a short time during the last two decades of the nineteenth century, but this 'railway town' that could send fish rapidly to the south Wales market was soon overshadowed by Milford. Between 1890 and 1895, for example, more fish were landed at Neyland than at Milford but by the end of the decade, Milford had outstripped its neighbour. In 1890 a total of 18,590 tons of fish were landed at Neyland and 5,610 tons at Milford. By 1899 a total of 14,375 tons were landed at Milford but 7,702 tons only at Neyland. By 1905 Neyland, once known as 'New Milford', had not only lost its steam packet service to Ireland but its fishing fleet as well. Today a marina dominates the old pier.

Aberystwyth, once the most important of Welsh fishing ports. The boat is a three-masted herring boat (Photo – R.J.H. Lloyd)

The Cardigan Bay ports

The fishing industry in the remainder of Pembrokeshire, in St Bride's Bay and northwards to the mouth of the Teifi, has always been overshadowed by Milford Haven, but inshore fishing, principally for lobsters, crabs, crawfish, queens and scallops as well as mackerel and pollack is carried out from a number of seaside villages such as Abereiddi, Little Haven and Porth-gain. The beautifully situated harbour at Solva supports about nineteen part-time fishermen mainly concerned with lobster and crab fishing off St David's Head and about ten inshore fishing vessels, principally for scallop fishing are based at Lower Fishguard. Scallops are landed at the port, together with minor quantities of lobster, mackerel, plaice, skates and whiting. Of course in the sixteenth and seventeenth centuries, Fishguard, located at the mouth of the River Gwaun in a very sheltered position, was a centre of the herring industry -the coast of north Pembrokeshire at that time was particularly rich in herrings. Salted and smoked herrings were exported in considerable quantities from Fishguard to other parts of Wales but more especially to Ireland; a country that was to figure prominently in the development of Fishguard down to the present day.

None of the Cardigan Bay ports north of St David's Head could ever boast of substantial fishing fleets and most of the boats engaged in the industry were small open boats concerned with inshore fishing. 'From Milford Haven to Holyhead' says a writer in 1884 'there is not a port into which a boat drawing eight feet six inches of water can run at all states of the tide.'[30] It was suggested that 'nothing but good and convenient harbours along the coast to make the present comparatively insignificant fisheries as important as those of any portion of the shores of Great Britain' would solve the problem.

In the early nineteenth century a number of harbours were constructed along the Cardigan Bay coast and although those harbours were built privately to provide port facilities for trading vessels they also provided a base for inshore fishing activities. The Aberaeron harbour constructed during the second decade of the nineteenth century was built to cope with the considerable coastal traffic that came to the port. A local newspaper pointed out that 'The coast abounds with fish, particularly herrings, in great abundance and few situations afford greater facilities for carrying on a Fishery on a considerable scale than this.'[31] Those dreams were never fully realized although Aberaeron has always possessed its small fleet of inshore vessels. The harbour at nearby New Quay although constructed principally for a merchant fleet in the 1820s nevertheless provided sheltered anchorage for a substantial inshore fishing fleet and half a dozen or more boats still operate from the port principally for the capture of mackerel, lobsters, crabs and scallops.[32]

In the seventeenth century, Aberystwyth replaced Aberdyfi as the principal hertIng port of Cardigan Bay and by 1755 the port supported sixty fishing vessels of 10 to 12 tons each, carrying seven men each during the autumn and winter fishing season:

> Here was an industry which could have brought wealth to the town, but there was no one prepared to do the necessary organising. In 1702, 1,734 barrels of fish were sent to Ireland in English or Irish ships. By the middle of the century the trade was almost entirely in the hands of merchants from Liverpool whose practice it was to send ships from the Isle of Man to Aberystwyth, there to purchase and salt fish ready for export.[33]

The Liverpool merchants made considerable profits from landings at Aberystwyth. By the end of the eighteenth century however, the town had declined as a fishing port although Hoylake, Fleetwood, Liverpool and Brixham vessels were able to catch substantial quantities off Aberystwyth. In the 1880s, Aberystwyth was said to possess a total of seventy-seven small fishing boats, but was still largely dependent on boats from Hoylake, Liverpool and Fleetwood.[34] After that date, the Aberystwyth fishing fleet declined and in 1928:

> ... there was only one first class motor boat. This boat is of the usual Cardigan Bay type i.e. thirty to forty feet long and eight to twelve feet beam – sailing boats but with marine engines ... The Aberystwyth boat is engaged in trawling throughout the year except for a period varying from two to three months in the early part of the year ... the grounds usually worked are those lying between Aberystwyth and New Quay. Within these limits the vessels may work on the fishing grounds extending from New Quay to Aberaeron or at the 'Gutter' – a soft clayey sea bed stretching from about five miles south of Aberystwyth to New Quay head – or on the more sandy bottom farther out beyond the Gutter, but never a greater distance than twelve miles from the shore.[35]

A total of about nine small rowing boats that fished close inshore from Aberystwyth also operated in 1928 being mainly concerned with the capture of whiting in the winter and mackerel in the summer.

Today minor quantities of shellfish, cod, plaice, whiting and mackerel are landed at Aberystwyth, the total in the 1980s averaging no more than 8.22 tons of whitefish and 9.66 tons of shellfish per year.

The estuary of the Dyfi was famous in the sixteenth and seventeenth centuries for being the meeting place of all the herring boats of the area for the commencement of the herring fishing season at Michaelmas. In the 1860s, there were great plans for making Aberdyfi a fishing port and depot in connection with the Cambrian Railway but those plans never materialized and the small port ceased to be 'a wonderfull greate resorte of ffyshers assembled from all places within the Realme'.[36] That port is of little consequence today.

North of the Dyfi estuary, although a great deal of fishing has been carried out down to the present day, the industry as such has always been small and fishing for a variety of sea fish, herrings in particular, has been carried out from small boats. Although Barmouth, for example, was one of the principal herring ports of Cardigan Bay in the Middle Ages, the fishing industry never developed to the extent that it could have done. Difficult communications above all discouraged the development of an inshore trawling industry and although a harbour was constructed at Barmouth in 1902, it was built specifically to export the products of Merioneth; woollen goods in particular.[37] In the early 1970s however, there was an expansion in fishing activities from Barmouth for not only did local fishermen take a greater interest in fishing activities, but the town attracted boats and men from other ports. Southern Ceredigion fishermen, for example, made Barmouth their home port and even Spanish fishermen were drawn to the area. The main reason for the modest revival of Barmouth as a fishing port was the discovery of valuable scallop and lobster beds in the northern part of Cardigan Bay. By 1976 the scallop beds had become almost exhausted due to overfishing by native and foreign boats and the revival of fishing activities that seemed possible in 1970 had come to nothing. In 1981 for example, the total quantity of fish landed at Barmouth was only 1,008 cwt. of which 740 cwt. were shellfish. In 1982 that low catch had been reduced to a total of only 876 cwt.

Gwynedd ports

Throughout Gwynedd, the fishing industry has by tradition been one pursued by many part-time fishermen operating in open boats from the small coastal villages of Llŷn and Anglesey. Most are engaged in activities ranging from salmon seine-netting to mackerel lining and from lobster potting to mussel dredging. Ports such as Porthmadog, Pwllheli, Caernarfon, Bangor, Conwy and Holyhead have their small fleets of trawlers and drifters and a variety of fish are landed at those ports. In May 1984 for example, the whole of north

The Conwy in-shore fishing fleet in the 1970s

Wales employed 122 full-time fishermen operating in 74 boats together with 151 part-time fishermen operating 147 boats.

Although in the past vast quantities of herrings, plaice and other fish were caught by Gwynedd fishermen, in recent years the most lucrative of activities has been the catching of lobsters and other shellfish. In the 1920s for example, Holyhead was of great importance in the herring fishing industry and Scottish drifters made it a practice to land their catches at the port. With first-class rail links to a lucrative market, Holyhead flourished as a fish port although it could only boast of fourteen fishing vessels itself. Today, the main activity of Holyhead fishermen is the capture of lobsters and crab off the rocky coasts of the west and north of Ynys Môn.

In 1987 Holyhead for the first time in its history overtook Milford in terms of cash turnover for fish. Fish processors became aware that dogfish not greatly appreciated by the British consumer could realize a considerable income from the European Continent, especially France. Holyhead is now the major dogfish port in the country. Holyhead Fish Processors, which started in 1982 with two long liner vessels, had shares in as many as forty boats that delivered fish to its processing plant in the port. In 1991 that promising venture faced bankruptcy.

The Dee estuary

Until the end of the nineteenth century Hoylake in the Dee estuary was of considerable importance as a fishing port, but the silting up of 'the Hoy lake' contributed to a decline of the industry. Hoylake boats travelled widely and they fished as far south as Cardigan Bay and occasionally in the Bristol Channel. Earlier in the nineteenth century 'before the introduction of the railways a fleet ... [they] used to come up the Dee every day from Flint and Hoylake to Chester with the morning tide bringing mackerel, turbot, whiting, soles and other fish according to the season. Chester fish market was an important institution in those days'.[39] The most important period for the Hoylake fishery was between 1885 and 1895. At Connah's Quay seine-netting for salmon still flourishes while at nearby Flint a number of trammel-netsmen drift for salmon and a variety of other fish.[40]

NOTES

Introduction

1. J.G. Jenkins, *Maritime Heritage – The Ships and Seamen of Southern Ceredigion* (Llandysul 1982) pp.6-10.
2. *The Times*, 24 October 1899.
3. *Pembrokeshire Herald*, 30 August 1912.
4. *Report of the Commissioners on Salmon Fisheries* (London 1841) p.16.
5. J.G. Jenkins, *Nets and Coracles* (Newton Abbot 1974) pp.lo8-9.
6. B.H. Malkin, *The Scenery, Antiquities and Biography of South Wales*, (1807) Vol. II, p.206.

Chapter 1

1. G. Owen, *The Description of Pembrokeshire* (1892 ed.) Vol. I, pp.117 *et seq.*
2. For a detailed study of commercial salmon and sewin fishing in Wales see J. Geraint Jenkins, *Nets and Coracles*, op.cit.
3. A. Rees, *The Cyclopaedia of Arts, Sciences and Literature* (1819), Vol. XXXI
4. Gerald de Barri *The Itinerary of Archbishop Baldwin through Wales in 1188* (1806 ed.) p.98.
5. Owen, op.cit. I p.117.
6. J.N. Taylor, *Severn Fishery Collection* (Gloucester 1953) p.20.
7. Colin Matheson, *Wales and the Sea Fisheries* (Cardiff, 1929) p.6.
8. Interview with Mr Hugh Jones, Aberdaron; Welsh Folk Museum Tape, No. 2780.

Chapter 2

1. Since the coracle is strictly a river craft not used in estuarine waters in Wales, it will not be considered in this book. For details see J.G. Jenkins, The Coracle (Carmarthen 1988)
2. Taylor, J.N. 'Elver Fishing on the River Severn' *Folk Life* III, pp.51 *et seq.*
3. Taylor, J.N., *Illustrated Guide to the Severn Fishery Collection* (Gloucester, 1953) p.13.
4. Davies, *General View of the Agriculture of South Wales* (London 1815) II p.66.
5. Owen, op.cit. II, p.117.
6. S. Lewis, *Topographical Dictionary of Wales* (1834) II, p.300.
7. D.C. Davies, 'The Fisheries of Wales', *National Eisteddfod Transactions* (Liverpool, 1885) p.300.
8. Ibid.
9. E. Lewis, 'The Goredi near Llanddewi Aberarth', *Archaeologia Cambrensis*, 1924, Series 7, Vol. 4 pp.395-8.

10. 24 & 25 Victoria CL109 28 & 29 Victoria CL121.
11. Taylor, op.cit. p.14.

Chapter 3

1. George Owen, *Description of Pembrokeshire* (1892 ed.) p.116.
2. C. Matheson *Wales and the Sea Fisheries* (Cardiff 1929) p.9.
3. Information from V.A. Lees, District Inspector, Wales Ministry of Agriculture, Fisheries and Food, Private communication, December 1974.
4. J. Pennant *British Zoology*, Vol. 3 (London, 1796) p.336-7
5. E.W. Knight-Jones, *Evidence in Milford Haven Tidal Barrage Bill* (1959).
6. Sam Jenkins, Clydfan, Aber-porth.
7. D. Carrellio Morgan, private correspondence.
8. Information from William Owen.
9. Information from William Evans.
10. E.A. Hall, *A Description of Caernarvonshire* (1805-1811) (Caemarfon, 1952) p.110.
11. MS. notes by J. Roberts, Nefyn.
12. Hall, op.cit., p.99.
13. Ibid., p.248.
14. J.F. Rees, *The Story of Milford* (Cardiff, 1954) p.110.
15. *Brut y Tywysogion* (1860 ed.) p.263.
16. Matheson, op.cit., p.15.
17. Ibid., p.15.
18. E.A. Lewis, *The Welsh Port Books 1550-1603* (Cardiff, 1927) pp.30 *et seq.*
19. Owen, op.cit., pp. 116-27.
20. Matheson, op.cit., p.25.
21. Ibid., p.27.
22. 23 Geo. II, c.24, 1750.
23. Pennant, op.cit., II, pp.386-7.
24. Hall, op.cit.
25. Herman Moll, *A New Description of England and Wales* (London, 1724) p.243.
26. H.P. Wyndham, *Tour through Monmouthshire and Wales in 1774 and 1777* (London, 1781).
27. E. Donovan, *Descriptive Excursions through Wales in 1804* (London, 1805).
28. S. Lewis, *Topographical Dictionary of Wales* (London, 1834) I, p.3.
29. Ibid., II, p.300.
30. Davies, D.C. 'The Fisheries of Wales' *Liverpool National Eisteddfod 1884 Transactions* (Liverpool 1885) pp.285-320.
31. Ibid., p.304.
32. Ibid., p.306.
33. D. Lleufer Thomas, *Digest of Welsh Commission Report* (London), p.28. .
34. *Report of the Superintendent, Lancashire and Western Sea Fisheries District* (Preston, 1994 ed.) p.120.
35. Matheson, op.cit., pAl.
36. Ibid., pAl.
37. Information from William Evans.
38. Based on stiitistics supplied by Lancashire and Western Sea Fisheries District, Preston.

39. Statistics produced by Ministry of Agriculture, Fisheries and Food.
40. R.J.H. Lloyd, 'Tenby Fishing Vessels', *Mariners' Mirror*, Vol. 44, No.2, May 1958.
41. Ibid., pp.104-5.
42. R.J.H. Lloyd, 'Aberystwyth Fishing Boats', *Mariners' Mirror*, Vol. 41.No.2, 1955, p.158.
43. Ibid., p.154.
44. Ibid., p.154.
45. Hall, op.cit.
46. University College of North Wales, Library Document 1680/1.
47. Lloyd, op.cit., p.153.
48. Ministry of Agriculture, Fisheries and Food, *Description of Principal Fishing Craft, Gear and Methods of Fishing* (London, n.d.) p.3.
49. Ibid.
50. Correspondence with R.W.K. Davies, District Inspector of Fisheries, Wales. March 1986.

Chapter 4

1. S.M.Tibbott, *Amser Bwyd* (Cardiff, Welsh Folk Museum, 1974) pp.66-8.
2. D.C.Davies 'The Fisheries of Wales', *Transactions of the Liverpool National Eisteddfod, 1884* (Liverpool, 1885), p.309.
3. J.M.Thomas, 'Cockles', *Gower: Journal of the Gower Society*, Vol. 2 (1949). p.32.
4. James Mason, *Mussels in Scotland*, Marine Laboratory, (Aberdeen 1973) p.3.
5. D.A.Hancock and A.E.Urquhart, *The Fishery for Cockles (Cardium, Edule L.) in the Burry Inlet, South Wales.* (HMSO Fishery Investigations Series II. Vol. XXV, Number 3 1966) p.1.
6. H. L. Bulstrode, *Shell fisheries other than oysters in relation to disease* (39th Report Government Board for Public Health. HMSO, 1911) p.194.
7. F.S. Wright, *Report on an inspection and survey of the Penclawdd and other cockle beds in connection with the taking of undersized cockles* (Swansea South Wales Sea Fisheries Committee, 1921) p.1.
8. Hancock and Urquhart, op.cit., p.21.
9. P.E. Davidson, *The Study of the Oystercatcher (Haematopus ostralegus -3E-.) in relation to the Fishery for Cockles (Cardium edule L.) in the Burry Inlet, South Wales.* Fishery Investigations series II, Vol. XXV, No.7, p.27 (HMSO, 1967).
10. *South Wales Evening Post*, 20 February 1975, 'The West Wales Cockle Industry is now in a critical position. The large flocks of cockle-eating oystercatchers is largely responsible for the poor landings ... During an oystercatcher cull which ended on December 6th last year, 9,238 birds were killed.'
11. C. Matheson, *Wales and the Sea Fisheries* (Cardiff, 1929) p.61.
12. Hancock and Urquhart, op.cit., p.25.
13. Hancock and Urquhart, op.cit., p.25.
14. David Davies of Penclawdd was still operating as a part-time basket-maker in 1973.
15. Information from Mrs Lettice Rees of Llansaint and Mrs Harriette Eynon of Crofty.

16. Hancock and Urquhart, op.cit., p.22.
17. Ibid., p.23.
18. Bulstrode, op.cit., p.194.
19. Hancock and Urquhart, op.cit., p.23.
20. In 1972 there were at least seven modern plants in Pen-clawdd owned by Mr Jim Howells, Mr R. Bird, Mr Stan Parkhouse, Mr Selwyn Jones, Mr Moelwyn Eynon, Mr Gurnos Rees and Mrs Coghlan. Some of them have subsequently closed, perhaps temporarily as a result of the decline in cockle gathering.
21. Thomas, op.cit., p.32.
22. Ibid., p.33.
23. *Book IV canto XI*, Verse xxxix.
24. R. Williams, *The History and Antiquities of the town of Aberconwy* (Denbigh, 1835) pp.144-5.
25. Davies, op.cit., p.295.
26. N. Tucker, *Conway and its Story* (Denbigh, 1960), p.124.
27. T. Jones, 'Perlau Tat' in *The Carmarthenshire Historian*, Vol. X (1973), pp.71-5.

Chapter 5

1. *Report of Departmental Commission on In-shore Fishing* (HMSO 1914)
2. W. Bingley, *A Tour Round North Wales* (1800) Vol. 1, p.470.
3. T. Pennant, *British Zoology* (1776) Vol. 3, p.336.
4. E,W.H. Holdsworth, *Deep Sea Fishing and Fishing Boats* (1874).
5. F. Buckland, and S. Walpole: *Report on the Crab and Lobster Fisheries of England and Wales, of Scotland and of Ireland* (HMSO 1877).
6. National Library of Wales, MS 10364E.
7. Davies., op.cit.
8. G. Owen, *The Description of Pembrokeshire* (1892 ed) Vol. I, p.126.
9. A. C. Simpson, *The Lobster Fishery of Wales*, Fishery Investigations, Series II, Vol. XXII, No.3 (1958) p.2.
10. Ibid.
11. Ibid. pp.28-33.
12. Statistics provided by Fisheries Division, Welsh Office.
13. F.M. Davies, *An Account of the Fishing Gear of England and Wales* (HMSO 1937) p.110.
14. Simpson, op.cit., p.8.
15. Interview with Mr Richard Williams, Welsh Folk Museum Tape 2795.

Chapter 6

1. Matheson, op. cit., p.50.
2. Lewis, E.A., *The Welsh Port Books* (London, 1927)pp.30 et seq.
3. Owen, G. op.cit.I, p.122.
4. Ibid.
5. Fenton, R. *A Historical Tour Through Pembrokeshire* (London, 1810) p.238.
6. D.C. Davies 'The Fisheries of Wales' *Liverpool National Eisteddfod Transactions* (1883) p.308
7. Fenton, op.cit. p.395.
8. R.J.H. Lloyd, 'Tenby Fishing Boats', Mariner's Mirror (1958), p.100.
10. Ibid.

11. Finch, Roger, 'The Tenby Lugger' in *Sailing Craft of the British Isles* (London 1976) pp.116-7.
12. Davies, op.cit. p.:no.
13. R.J.H. Lloyd, 'The Mumbles Oyster Skiffs', *Mariner's Mirror* Vol. 40 (1954) p.260.
14. E.W.H. Holdsworth, *Deep Sea Fishing and Fishing Boats* (1874) pp.184-5.
15. Lloyd, op.cit., p.263.
16. J.H. Wade *Glamorganshire* (1914) p.97.
17. F.S. Wright, *Fishery Investigations*, Series II, Vol. XII, No.4, (HMSO 1932).
18. *Swansea Guide* (1880).
19. Lloyd, op.cit. pp.260-1.
20. Ibid, p.262.
21. Wright, op.cit.
22. Ibid, p.261.
23. J.G. Jenkins, *Nets and Coracles* (Newton Abbot, 1974) p.37.
24. Davies, op.cit. p.306.
25. J. Mason *The Scottish Fishery for Scallops and Queens* (HMSO 1972) p.5.

Chapter 7

1. Matheson, op.cit. p.6.
2. Owen, G. *Description of Pembrokeshire* (1892 ed.) VoU, p.123.
3. Matheson, op.cit. p.6.
4. Ibid.
5. A.J. Lees, *The Local Fishing Industry*, unpublished MS.
6. Davis, op.cit. p.30.
7. Ibid., p.35.
8. *Report of the Commissioners on Salmon Fisheries* (1861) p.15.
9. Ibid., p.35.
10. Ibid., p.210.
11. Ibid., p.xi.
12. An example dating from 1931 is preserved at the Welsh Folk Museum, St Fagans.
13. M.e. Brown, 'Mud Horse Fishing in Bridgewater Bay', *Folk Life*, Vol. 18 (1980) pp.24-6.
14. Ibid.
15. Matheson, op.cit. p.68.
16. *Argae*, pl. *argaeau*, is derived from the prefix *ar* + *cae* (enclosure); *cored*, pl. *coredau*, is derived from *cor* (plaiting or binding) + suffix ed. compare Old Breton -*coret*; Irish -*cora*.
17. Davis, op.cit., p.1.
18. *Report of the Commissioners Appointed to Inquire into Salmon Fisheries* (England and Wales). (London HMSO 1861), pp.ix-xi.
19. Ibid.
20. E. Lewes, 'The Goredi near Llanddewi Aberarth', Arch. Camb. Vol. LXXIX (1924) p.399.
21. Ibid., p.398.
22. D.C. Davies, 'The Fisheries of Wales', *Liverpool National Eisteddfod Transactions 1884* (Liverpool 1885) p.309.
23. Matheson, op.cit., 13.

24. H.R. Davies, *The Conway and Menai Ferries* (Cardiff 1942) p. 282.
25. J. Evans, *Letters written during a tour through North Wales* (London, 1804) p.231.
26. D. Senogles, *Ynys Gored Goch* (Menai Bridge, 1969).
27. J.G. Jenkins, *Nets and Coracles* (Newton Abbot 1974) pp.206-16.
28. Ibid., pp.217-22.
29. Davis, op.cit. p.8.
30. Ministry of Agriculture, Fisheries and Food. Information Sheet L.2 (n.d.).
31. Burgess, J. *Otter Trawling* (Bridport, n.d.) pp.5-6.
32. Matheson, op.cit. p.8.

Chapter 8

1. Jenkins, *op.cit.*, *Nets and Coracles* p. 255
2. I am grateful to Mr J.K. Neale of Truro, grandson of the founder of the company for a great deal of information on the company.
3. *Active* (55 gross tons), *Alert* (57), *Brisk* (56), *Campania* (67), *Champion* (58), *City of Aberdeen* (65), *Duke of York* (58), *Euphrates* (68), *Federal* (67), *George* (58), *Grassholm* (40); *Labore et Honore* (58), *Lark* (52), *Luciano* (73), *Monarch* (65), *Nemophila* (84), *Pride of the Humber* (58), *Skomer* (40), *St Laurence* (75).
4. Matheson, op.cit. p.81.
5. Mr J.K. Neale, private correspondence.
6. *Fuji* (255 gross tons), *Hatano* (255), *Hirose* (227), *Ijuin* (257), *Kudama* (257), *Kuroki* (248), *Mikasa* (255), *Miura* 257), *Nodzu* (257), *Nogi* (257), *Oku* (248), *Oyama* (257), *Sasebo* (265), *Saxon* (239), *Settsu* (231), *The Norman* (225), *The Roman* (224).
7. *Asama* (built 1919), *Hatsuse* (1920), *Kumiski* (1917), *Mikasa* (1915), *Miura* (1916), *Suma* (1924), *Tamura* (1917), *Fuji* (1912), *Hatano* (1912), *Kodama* (1911), *Kuroki* (1911), *Nogi* (1908), *Oku* (1909), *Oyama* (1908) and *Settsu* (1912) of the pre-war fleet were returned by the Admiralty to Neale & West Ltd. after war service.
8. Matheson, op.cit. p.83.
9. In 1939 the fleet consisted of 14 vessels: *Akita* (300 gross tons, built 1939), *Asama* (303 tons, b.1929), *Hatsuse* (295 tons, b.1927), *Honjo* (308 tons, b.1928), *Kunishi* (303 tons, b.1927), *Muroto* (340 tons, b.1931), *Naniwa* (340 tons, b.1931), *Nodzu* (303 tons, b.1929), *Oku* (303 tons, b.1929), *Oyama* (340 tons, b. 1931), *Sasebo* (308 tons, b.1928), *Sata* (340 tons, b.1931), *Suma* (303 tons, b.1927), *Yashima* (303 tons, b.1929).
10. In 1950 the fleet consisted of the following vessels: *Aby* (built 1945), *Akita* (1939), *Braemar* (1927), *Fort Dee* (1929), *Hatsuse* (1927), *Muroto* (1931), *Nodzu* (1929), *Oku* (1929), *St Botolph* (1946), *Sasebo* (1928), *Sata* (1931), *Warlord* (1914), *Yashima* (1929).
11. *Sea Breezes*, July 1956, p.11.
12. W.H. Jones *History of the Port of Swansea* (Carmarthen 1922) p.75.
13. The fleet consisted of the following vessels: *Amroth Castle*, *Benton Castle*, *Carew Castle*, *Dale Castle*, *Hariat Castle*, *Hene Castle*, *Lawrenny Castle*, *Manorbier Castle*, *Narberth Castle*, *Pembroke Castle*, *Picton Castle*, *Points Castle*, *Roche Castle*, *Tenby Castle*, *Upton Castle*, *Valmyris Castle*.

14. *Amelia* (built 1906), *Izaak Walton* (1907), *Cardiff Castle* (1907), *Carew Castle* (1912), *Clyne Castle* (1907), *Dale Castle* (1909), *Hene Castle* (1909), *Kidwelly Castle* (1907), *Lawrenny Castle* (1908), *Neath Castle* (1913), *Oxwich Castle* (1907), *Pennard Castle* (1907), *Penrice Castle* (1913), *Picton Castle* (1911), *Roche Castle* (1910), *Swansea Castle* (1912), *Tenby Castle* (1908), *Langland* (1905), *Caswell* (1908).

15. Jones, p.325.

16. Matheson, p.79.

17. Ibid. p.80.

18. Margaret Davies, *The Story of Tenby* (Tenby 1979), p.15.

19. R.J.H. Lloyd, 'Tenby Fishing Boats' in *Mariner's Mirror* (1954), p.94.

20. Ibid., p.98.

21. Ibid., p.98.

22. J.F. Rees, *The Story of Milford* (Cardiff 1954) p.91.

23. Ibid., p.103.

24. Ibid., p.lo6.

25. Matheson, op.cit. p.77.

26. In 1926 David Pettit owned 13 vessels: *Caliph, Cleopatra, Cyelsi, Thomas Deas, Calydonic, Capstone, Cawdor, Cheriton, Clyro, Cornelia, Cotsmuir, Cresswell, Undea.*

27. In 1928 Brand and Curzon owned 20 vessels: *Bardolph, Caliban, Charles Boyes, Cicero, Isaac Heath, James Aldridge, James Archibald, John Baptish, Joseph Barrett, Joseph Button, Phineas Beard, Robert Bowen, Robert Gibson, Thomas Bartlett, Thomas Booth, Thomas Bryan, Thomas Connolly, Morgan Jones, William Bell.*

28. *British Trawler Federation Annual Report 1964* (London, 1965) p.38.

29. *Bryher* (built 1961), *Constant Star* (1962), *Norrard Star* (1958), *Picton Sea Eagle* (1958). *Picton Sea Lion* (1956).

30. Davies, D.G. 'The Fisheries of Wales', *Transactions of the LiverpooL National Eisteddfod 1884* (Liverpool 1885) p.285.

31. *Cambrian Journal*, 17 August 1911.

32. J.G. Jenkins, *Maritime Heritage* (Llandysul, 1982) pp.65-7.

33. W.J.Lewis, *Born on a Perilous Shore* (Aberystwyth 1980) p.93

34. Davies, op.cit. p.286.

35. Matheson, op.cit. p.41.

36. E.A. Lewis, *The Welsh Port Books* (London, 1927) p.309.

37. Lloyd, L. *Maritime Merioneth -The Town and Port of Barmouth* (Porthmadog 1974) p.18.

38. Statistics provided by the Lancashire & Western Sea Fisheries Committee.

39. Matheson, op.cit. p.35.

40. Jenkins, Nets and Coracles op.cit., pp.217-22; 256-7. 161

INDEX